72012

REF QE 905 .C55 1999

Cleal, Christopher J., 1951-

Plant fossils

REFERENCE
FOR LIBRARY USE ONLY

FOSSILS ILLUSTRATED
Volume 3

PLANT FOSSILS

FOSSILS ILLUSTRATED

ISSN 0960–8664

Series Editors
Douglas Palmer and Barrie Rickards

Volume 1
GRAPTOLITES: WRITING IN THE ROCKS
edited by
Douglas Palmer and Barrie Rickards

Volume 2
TRILOBITES
H. B. Whittington

PLANT FOSSILS

THE HISTORY OF LAND VEGETATION

Christopher J. Cleal
Barry A. Thomas

THE BOYDELL PRESS

© Christopher J. Cleal and Barry A. Thomas 1999

All Rights Reserved. Except as permitted under current legislation no part of this work may be photocopied, stored in a retrieval system, published, performed in public, adapted, broadcast, transmitted, recorded or reproduced in any form or by any means, without the prior permission of the copyright owner

First published 1999
The Boydell Press, Woodbridge

ISBN 0 85115 684 3

The Boydell Press is an imprint of Boydell & Brewer Ltd
PO Box 9, Woodbridge, Suffolk IP12 3DF, UK
and of Boydell & Brewer Inc.
PO Box 41026, Rochester, NY 14604–4126, USA
website: http://www.boydell.co.uk

A catalogue record for this book is available
from the British Library

Library of Congress Cataloging-in-Publication Data
Cleal, Christopher J.
  Plant fossils : the history of land vegetation / Christopher J. Cleal, Barry A. Thomas.
    p.   cm. – (Fossils illustrated, ISSN 0960–8664 ; v. 3)
  Includes bibliographical references and index.
  ISBN 0–85115–684–3 (alk. paper)
  1. Paleobotany. 2. Plants, Fossil. I. Thomas, Barry A. II. Title. III. Series.
  QE905.C55   1999
  561–dc21   99–19437

This publication is printed on acid-free paper

Printed in Great Britain by
St Edmundsbury Press Ltd, Bury St Edmunds, Suffolk

CONTENTS

General Editor's Preface		vii
Acknowledgements		ix
Chapter One	Introduction	1
Chapter Two	Early land plants	11
Chapter Three	Club-mosses	27
Chapter Four	Horsetails	40
Chapter Five	Ferns	50
Chapter Six	Early seed plants	62
Chapter Seven	Modern seed plants	82
Chapter Eight	Flowering plants	98
Chapter Nine	History of land vegetation	108
Chapter Ten	Highlights of palaeobotanical study	120
Appendix 1	Classification of vascular plants	138
Appendix 2	Further reading	147
Explanations of Plates 1–128		156
Index		180
Plates		189

GENERAL EDITOR'S PREFACE

Without plants, life on Earth, as we know it, would not exist. Plants have played a central role in the evolution of life and the colonisation of land. And yet the fossil history of plants is not well known, even to many scientists, beyond a general knowledge that coal is formed of fossil plant material. The evolution of the Earth's flora has had its vicissitudes just as much as that of the fauna. New groups of plants have come and gone, there have been radiations and extinctions.

Overall there has been a major change in the domination of plant life from the evolution of primitive vascular plants some 400 million years ago. These first land plants were tiny leafless stems which only grew upright for a few centimeters and were restricted to lowlying waterlogged habitats. Cleal and Thomas tell the remarkable story of how, from these unpromising beginnings, the whole of Earth's flora has evolved. Each of the major groups of plants is described in general order of appearance in the record. The history develops through the first giant club-mosses, horsetails and ferns, which contributed so much to the first forests on Earth and the economically famous coal deposits of Carboniferous times. By the time the dinosaurs took over, in the Mesozoic Era, landscapes were dominated by coniferophytes and pteridophytes. In turn these were displaced as the angiosperms and large plant eating mammals finally burst into dominance in the Tertiary some 60 million years ago.

'Plant Fossils' tells this remarkable 400 million year history of land vegetation and its photographically illustrates it with a generous selection of fossil plant portraits. Most of these photographs of fossil plants have never been seen before, outside of academic journals and represent an international sample of the plant record. The authors also tell of the scientists who have contributed to the development of this story and hazard some predictions about the directions which future research may take.

As professional palaeobotanists, Chris Cleal and Barry Thomas have made life-time studies of ancient flora and are ideally placed to tell this remarkable story. They are well known researchers (Dr Cleal at the National Museums and Galleries of Wales, Cardiff and Professor Thomas at the University of Wales Lampeter) within the very active international community of palaeobotanists and consequently are able to present the most up-to-date view and interpretation of the fossil record. The authors have benefitted from the cooperation of their colleagues around the world, especially those in North America, who have generously provided many of the photographs. The book reflects the truly international story of fossil plants.

The book is the third in a series called 'Fossils Illustrated' to be published. The overall aim is to allow professional palaeontologists to tell their particular fossil 'story' as they think best and to illustrate it with as many photographs as possible. I would like to thank the publishers, The Boydell Press, for giving us the original opportunity to present these fossil 'stories' and continuing to support the series.

Douglas Palmer,
series editor,
Cambridge 1999.

ACKNOWLEDGEMENTS

This book is a testament to the fraternity of the palaeobotanical community; without the provision of many of the photographs by our colleagues, it would not have been possible to produce this volume. Thanks go to (in alphabetic order): Dr Heidi Anderson (National Botanical Institute, Pretoria), Mr Y. Arremo (Swedish Museum of Natural History, Stockholm), Dr Sidney Ash (Albuqurque, New Mexico), Dr Richard Bateman (Royal Botanic Garden, Edinburgh), Professor David Batten (University of Wales, Aberystwyth), Professor Bill Chaloner (Royal Holloway and Bedford New College, University of London), Dr Margaret Collinson (Royal Holloway and Bedford New College, University of London), Professor David Dilcher (Florida Museum of Natural History, Gainsville), Dr Bill DiMichele (Smithsonian Institution, Washington DC), Mr Neil Ellis (Joint Nature Conservation Committee, Peterborough), Dr Muriel Fairon-Demaret (Université de Liège), Dr Else Marie Friis (Swedish Museum of Natural History, Stockholm), Dr Pat Gensel (University of North Carolina), Dr Alan Hemsley (University of Wales, Cardiff), Dr Paul Kenrick (Natural History Museum, London), Professor Hans Kerp (Westfälische Wilhelms-Universität, Münster), Dr Franz-Josef Lindeman (Paleontologisk Museum, Oslo), Professor Jean-Pierre Laveine (Université de Lille), Dr Steve Manchester (Florida Museum of Natural History, Gainsville), Dr Nick Rowe (Université de Languedoc, Montpellier), Professor Wang Ziqiang (Chinese Academy of Geological Sciences, Tianjin), Dr Joan Watson (University of Manchester) and Dr Erwin Zodrow (University College of Cape Breton, Sydney NS). Thanks also go to Dr John Faithful (Hunterian Museum, Glasgow) for the loan of one of the slides from the Kidston Collection, which we show on Plate 17. Where the images were produced from negatives supplied by colleagues, the final prints were prepared by staff at the National Museums and Galleries of Wales, Cardiff: Dr D. M. Spillard (Department of Biodiversity and Systematic Biology) and Mr Jim Wild (Photography Department). The illustrated specimens from the collections in the National Museums and Galleries of Wales were photographed by the museum's Photography Department, to whom we are deeply grateful. We thank Professor David Dilcher for checking the geological ranges of the angiosperm families given in Appendix 1. The text has greatly benefited from improvements suggested by Dr Douglas Palmer (Cambridge), who originally invited us to write this book. Finally, we would like to acknowledge the great skill of Mrs Annette Townsend (Cardiff), Dr Deborah Spillards and Mrs Pauline Dean (Guildford), who prepared the models and drawings of plant fossils illustrated in the book.

CHAPTER ONE

INTRODUCTION

The popular image of a 'typical' fossil is that of animal remains – a seashell or a dinosaur bone. There are indeed vast numbers of animal fossils, but there are also innumerable plant fossils. Plants first appeared on land about 425 million years ago, and since that time their remains have found their way into muds, silts and sands, to be preserved as fossils. These plant fossils give us our only direct view of the vegetation of the past, and from this we can infer the history of the evolution of plants and floras.

This is neither a palaeobotanical textbook nor an identification guide – there are many other books that fulfil these roles (see Appendix 2). Our aim instead is to illustrate the sort of plant fossils that palaeobotanists deal with and the type of information that can be gleaned from them. The coverage will reflect the relative abundances of different types of plant in the fossil record, rather than their original abundance in past vegetation. It certainly does not reflect the abundance of plant-types today.

WHAT IS A PLANT?

Early naturalists divided living organisms into two kingdoms, animals and plants. Animals were thought of as mobile organisms whose nutrition was based on the consumption of other organisms, whereas plants were static, green organisms whose food was generated by internal processes (mainly photosynthesis). Although this was satisfactory for classifying most of the organisms that we meet on a day-to-day basis, it soon became evident that the distinction was not clear-cut. Fungi were the most obvious major group that did not fit the twofold division, being static but dependent on other organisms for food, and these eventually were assigned to their own kingdom. As biologists looked more closely at the microscopic world, the position became even more complex.

In any system of classification, plants cannot be defined just by their photosynthetic nutrition, which is not universal throughout the kingdom (broomrapes and toothworts are not photosynthetic) and is also shared with certain bacteria (simple unicellular organisms without nuclei, such as blue-green 'algae') and protoctists

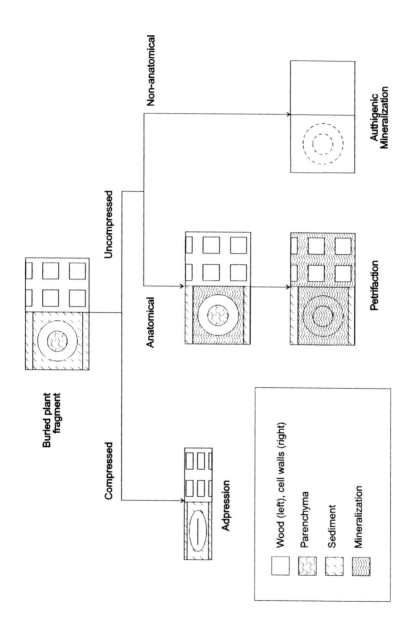

Text-figure 1. Summary of modes of preservation. After an original by Richard Bateman.

(enigmatic 'lower' organisms that have nucleate cells but which do not fit in any of the other groups – the flagellated alga *Chlamydomonas*, for instance). All plants are sessile, but this can hardly be taken as a defining character; fungi, and some protoctists and animals are also sessile. Plants are instead defined as those organisms that have alternating sexual (gametophyte) and asexual (sporophyte) generations, and where sexual activity produces an embryo on the gametophyte. Some algal protoctists have alternating sexual and asexual generations, but they do not produce embryos. Animals produce embryos but of a fundamentally different type, consisting of a hollow ball of cells that is usually detached from the tissue of the mother; the plant embryo, in contrast, is a solid structure that remains embedded in the maternal tissue. Animals also differ from plants in not having alternating generations (discussed further in Chapter 2). Fungi have neither embryos nor alternating generations.

In this account we are restricting ourselves to bryophytes and vascular plants, together with some primitive non-vascular plants found in the early phases of land plant evolution in the Silurian and Devonian. The bryophytes are the group of land plants that we know today mainly as mosses, hornworts and liverworts. Although perfectly adapted to life on land, they have not been able to develop into large organisms, as have the vascular plants. Vascular plants include most land vegetation such as ferns, horsetails, club mosses, and seed- and flowering-plants. Their main defining feature is a stem with vascular tissue, which assists in supporting the plant and in the transport of nutrients and water around its body. We are excluding the algae because, with a few notable exceptions, such as those that produce resistant cysts (e.g. dinoflagellates) or have a mineralized 'skeleton' (e.g. dasyclads), the fossilization potential of algae is very low and we know relatively little about their evolutionary history. This is in contrast to the true land plants, whose resistant structures (cuticles, vascular tissue, etc.) have provided a much more complete fossil record.

TYPES OF PLANT FOSSIL

A plant fossil can be formed when part of a plant finds its way into sediment or some other preserving medium (e.g. amber). Only very rarely are whole plants preserved as fossils. The subsequent changes that occur to that fragment control the type of information that the fossil will eventually yield (Text-fig. 1).

The commonest type of fossil is known as an adpression and most of the fossils illustrated in this book are of this type. A adpression is produced when the build-up of additional sediment on top of the plant fragment causes it to become vertically compressed. The pressure causes the plant tissue to be flattened into a thin layer of coal called a phytoleim. If the phytoleim is still preserved, the fossil is known as a compression, but if the layer is lost either though geological changes (e.g. additional compression and heat) or weathering after the fossil has become exposed, it is known as an impression.

Adpressions can retain the shape and some surface details of the plant fragment, especially of organs that originally had a flattened form, such as leaves. Simply splitting the rock will reveal such fossils. They rarely need much additional preparation, as the flat plant fragment will normally have settled in the sediment parallel to what would later become the bedding plane. A little small-scale excavation with fine needles or a small chisel (a process sometimes called dégagement) is usually all that is required. More three-dimensional organs, especially non-rigid ones such as flowers, can be significantly distorted by the compressive forces. There have been some attempts to decompress such compressions but these have met with only limited success. The fracture plane of the split rock usually passes over the smoother surface of the plant fossil and usually results in originally attached structures such as reproductive organs remaining embedded in the rock. The fossil can be separated from the rock by mounting it on a glass slide or in clear resin, and dissolving the rock away in hydrofluoric acid. This allows the compression to be observed from both sides, which can be of great help in the interpretation of many specimens.

If the sediment around the plant fragment hardens before significant compression can take place, the three-dimensional form of the fragment may be partly preserved. For example, in certain types of sediment, mineral growth such as siderite (iron carbonate) may occur quickly surround a plant fragment to form an encapsulating nodule (an example is shown on Pl. 99). The plant tissue itself will at least partly decay, but the resulting cavities in the nodule (or a subsequent growth of mineral in the cavities to form a cast) can reveal considerable detail of the plant. Such fossils are referred to as authigenic mineralizations. They may be studied directly, similar to adpressions, or it may be better to take a latex cast to produce a solid replica of the original plant fragment. Authigenic mineralizations can show the very fine surface detail of the fossil, even the individual cells, that can be studied by scanning electron microscopy. Sometimes authigenic mineralizations of fructifications still contain spores, which can be prepared and studied microscopically.

Moulds and casts of robust parts of plants such as wood and some seeds can be formed if the sediment itself is less prone to compaction, such as sand. After the sediment has hardened and the plant tissue decayed, the resulting cavity is known as a mould. If the mould becomes filled with sediment or mineral, a cast is formed (e.g. Pl. 23).

Another type of cast occurs where stems have a central core that is either hollow or made of soft, easily decayed cellular tissue. If such a stem falls into water, sediment may fill the central cavity before the rest of the tissue has decayed. After the surrounding tissue has decayed, there remains the sediment-cast of the internal cavity, referred to as a pith cast (e.g. Pl. 53). Horsetails and cordaites are particularly prone to this mode of preservation (see Chapters 4 and 7).

Most fossils preserve the shape of the original plant fragment, and may include the fine details of structures such as sporangial bodies. Palaeobotanists are always seeking such details of the anatomy of the plants as they can provide much clearer evidence of the plant's affinities. One of the most commonly preserved sources of anatomy in plant fossils is the cuticle. This outer 'skin', that most land plants possess, can reveal details of the immediately underlying epidermal cells, including diagnostic breathing pores (stomata), hairs and glands. To prepare the cuticle, the fossil is separated by removing the rock with hydrofluoric acid and then dissolving the coalified plant tissue with a strong oxidizing solution, such as Schulze's Solution (a mixture of nitric acid and potassium chlorate). The cuticle is then cleared in an alkali solution and finally mounted for observation either by light or electron microscopy. Examples of fossil cuticles are shown on Pls 19 and 111. The outer parts of spores and pollen are formed of sporopollenin, which may also be resistant to decay, and can be prepared and studied in the same way as cuticle (e.g. Pl. 34).

The fossils which show the best anatomical detail are petrifactions. These are formed when fluids containing minerals (e.g. calcite, silica) in solution have percolated through the body of the plant before significant decay and compression has occurred. The cells themselves become impregnated by the mineral, thus preserving their form. The cell wall is sometimes retained as a thin layer of coal around the mineral replacement of the cell contents (cell lumen). Alternatively, the cell wall is itself replaced by mineral. Either way, the detailed anatomy of the plant fragment is revealed when a section is cut through the fossil. Examples of such petrifactions can be seen on Plates 1–2, 17–18, 24–25, among others.

Traditionally, such fossils were studied using ground thin sections, similar to those used for the study of rock petrology. Where the cell walls are still retained as coalified layers, however, the 'peel' method may produce better results. A flat surface of the fossil is polished and then briefly etched in acid to remove a thin layer of the mineral. This projecting coalified cell walls are flooded in acetone and a thin sheet of acetate laid over the surface. The acetate embedded cell walls can then be peeled away from the fossil. The acetate film thus has preserved in it a thin section through the fossil, which can be examined under the microscope, revealing remarkable detail of the anatomy of the plant tissue.

Petrifactions only occur under unusual conditions, such as habitats associated with volcanic activity or where the plant fragments have been soaked in sea water. They are therefore much rarer than adpressions and casts. The information obtained from petrifactions must therefore be seen in the wider context of the adpression/cast record, and integrated with the entire plant fossil record to obtain as wide a picture as possible of past plant life.

WHERE ARE PLANT FOSSILS FOUND?

Plant remains can be found fossilized in most types of sedimentary rock. However, the most abundant and best preserved tend to occur in rocks that were formed in non-marine environments. Rocks formed from sediment deposited in river deltas provide some of the best opportunities to find plant fossils, especially the finer-grained sediment deposited in lakes within the deltas. If the water table remains generally high in the delta sediment, plant fragments will be much slower to decay and thus stand a greater chance of being entrapped within sediment. If sedimentation is very slow, peat will build-up and over geological time this can result in coal. If more sediment is being deposited, the plant fragments will be buried in sandstones or mudstones, becoming adpressions or casts, as described above. Both situations may occur in repeated sedimentary successions, resulting in seams of coal separated by sandstones and mudstones containing plant fossils. The Late Carboniferous coal-bearing sequences ('Coal Measures') of Europe and eastern North America are prime examples of this.

If an area in which peat has been formed becomes flooded by sea water, mineral deposits such as calcite nodules can form within the peat. If this occurs before the plant fragment has decayed significantly, the mineral can impregnate the plant tissue and crystallize within the cells, to form petrifactions (as described in the last section). The best known examples of such preservation are the coal balls from the Late Carboniferous of Europe and North America.

If the water table is lower, plant decay is much quicker and peat tends not to build up. Low water table conditions can be recognized by the oxidization of the sediment which results in red-coloured rocks. Plant fragments may still become encapsulated in the sediment in such situations, but on the whole they are less common than when the water table was higher.

Volcanic landscapes tend to be very unstable and their sediment deposits suffer from considerable reworking. On the face of it, this would seem to be not the most likely situation to find plant fossils. However, the ground-waters in such environments are often rich in minerals, which can petrify plant-fragments. An example of this is the famous Rhynie Chert deposit, formed when an Early Devonian peat accumulation was inundated by mineral rich waters, preserving in exquisite detail the internal anatomy of the very early land plants (see Chapter 2).

Lagoonal and coast deposits may contain fragments of plants that had drifted from the coastal vegetation. An example here is the Jurassic Stonesfield 'Slate' flora of Oxfordshire, where plant fossils including conifer and fern fragments are found with the remains of marine animals such as bivalves. Another is the Sheppey Flora from Kent, which has Tertiary plant remains associated with shark teeth and crab remains. The diversity and preservation of such drifted floras is sometimes poor (although this cannot be said for the Sheppey Flora) but they may preserve remains of vegetation that is different from that found in deltaic deposits.

HOW DO PLANT FRAGMENTS GET INTO THE FOSSIL RECORD?

Very occasionally, whole plants get entrapped in sediment and preserved where they originally grew, but more normally we see preserved fragments (leaves, seed, branches, etc.) that have been detached from the plant. The process of fragmentation may be part of the natural strategy of the plant, such as the deciduous shedding of leaves in many angiosperm trees growing in temperate climates. More often, however, it was the result of traumatic external influences, such as storm damage.

Once detached from the plant, the fragment may fall to the ground near where the parent plant was growing, but more often than not this will have been an exposed area where the plant tissue would rapidly degrade. Preservation is more likely where the plant is transported away from the place of growth usually by air (wind) or water (rivers), to where it can be rapidly covered by sediment, such as in a lake.

As a consequence of this, most plant fossil assemblages do not normally represent a single natural assemblage of plants, but a mixture of plant fragments from different vegetational habitats. Reconstructions of original vegetation can only be directly achieved in those rare situations where the plant remains are preserved more or less *in situ*, such as the 'fossil forests' of tree stumps or where the landscape has been subjected to a sudden catastrophic inundation by sediment (Pl. 26). The fossil record can in fact give a very biased view of the original vegetation. For instance, the Late Carboniferous Coal Measures, where most of the fossils found in the mudstones and siltstones represent the very narrow band of vegetation which grew along the river banks. The vast bulk of the forests were dominated by giant club-mosses growing in the back-swamps which are not preserved as adpressions nearly as often as the riparian plants. Much of what we know about the composition of the back-swamp vegetation comes from looking at plant fragments in coal seams (spores or petrified specimens from coal balls) which are the remains of peat produced on the forest floor.

Another consequence of the fossilization process is that most plant fossils are fragments of the original plant. In the case of some of the larger organs, such as fronds of ferns or Palaeozoic gymnosperms, even the organs have been fragmented. Reconstructions of plants depend on piecing together the chance finds of attached organs, such as a leaf and seed, a leaf and a stem, and a stem and a cone. Mostly, we know nothing of the seeds that were borne by a the plant that produced a particular leaf, let alone the appearance of the whole plant. This has consequence for naming plant fossils, which we will come to later.

BIAS IN THE FOSSIL RECORD

The fossil record provides a very incomplete picture of past vegetation because fossils can only be preserved where sediment accumulates and stands a good chance of not being reworked by erosion. The plant record is probably at its best in the very early phases of their migration onto land, in Silurian and Early Devonian times, as the wet habitats being colonized were ideal for fossilization. As soon as plants had adapted to drier habitats, bias in the record becomes significant. Only those plants that continued to occupy the wetter habitats, such as river banks and deltas and lake shores, are well represented as fossils. Since many of the main evolutionary developments in plants have been adaptations to life in drier regimes (especially reproductive adaptations such as seeds and flowers – see Chapters 6–8), there is a very real problem in using the fossil record directly to develop models that explain the evolutionary history of plants.

Is there any direct evidence of this incompleteness of the record? There are two main types of such evidence. Firstly, there is the fossil record itself. Sedimentary rocks sometimes yield tiny fragments of plants that have travelled a much longer distance than the larger plant fossils that they accompany. These may represent the vegetation growing in more elevated and thus drier habitats. Spores and pollen can travel vast distances, although the problem here is often a matter of discerning which group of plants a particular spore or pollen came from. The recent discovery of fragments of charcoal (technically called fusain) in lowland sediments provides us with further evidence. Charcoal, which is the product of fire, can preserve surprisingly fine detail of the plant, including some internal anatomy. Because of this anatomical detail, it is often possible to identify quite small pieces of charcoal. For example, fragments of charcoalified conifer leaves in coal-bearing strata in Yorkshire are some 5 million years older than the oldest conifer remains from the more usual fossil record. It is clear that while the lowland vegetation was dominated by spore-bearing plants and various now-extinct types of seed-plant, there were some conifer forests in the uplands.

There are also indirect means of judging the incompleteness of the plant fossil record. One example is known as 'the molecular clock'. By determining the rate at which genes randomly mutate, it is possible to use the number of genetic differences between two organisms to estimate how long ago they shared a common ancestor. Such evidence suggests that the flowering plants diverged from the other seed-bearing plants as long ago as Triassic times (if not earlier). But this is in marked contrast to the evidence of the fossil record, in which the oldest unequivocal occurrence of flowering plants is in rocks of early Cretaceous age. So, either the molecular clock has been 'running fast', or there were flowering plants growing in upland areas for millions of years before they appeared in the lowland habitats which are sampled by the fossil record.

NAMING PLANT FOSSILS

As with most aspects of natural science, naming specimens is a crucial aspect of palaeobotany. Without an accurate way of recording exactly what is found, there is no way that the subject can develop beyond simple descriptions of individual objects. It is normal to name plant fossils using a system of nomenclature very similar to that used in botany; they are, after all, the remains of once living plants.

This is not the place to go into a detailed discussion of botanical nomenclature, because there are several clear accounts of this subject (see bibliography at the end of the book). However, there is one important point where palaeobotanical nomenclature differs from botanical nomenclature and this can cause confusion. Rarely can a plant fossil be named in the same way as one would a living plant, because most fossils remains are only fragments of the original plant and so the whole organism is not available. Palaeobotanists have had to resort to a system of nomenclature whereby isolated parts of the plant are given separate species and genus names. One of the most popular examples used to explain this is the Late Palaeozoic giant club-mosses. Different parts of the plant are assigned to different genera (in this context they are called form-genera): *Stigmaria* is used for the rooting structures, *Lepidodendron* for the stems, etc. Within each form-genus, species are based on characters of the relevant organ, rather than estimates of what the original, natural species groups would have been if they had been based on whole plants. The rooting structures, for instance, are virtually indistinguishable in most of the giant club-mosses and tend to be assigned to a single form-species, *Stigmaria ficoides*. Form-species of stem are based mainly on the more variable characters of the leaf-cushions. Many different stem species have been recognized but some may merely represent variation within a natural species or even variation between different parts of the plant. Palaeobotanists normally strive to make the form-taxa as natural as possible, by incorporating features such as epidermal structure from the cuticles, but it is rarely possible to be certain that a form-taxon correlates exactly with a natural, whole-plant taxon.

Different types of preservation can also give rise to different form-generic and -specific names. Each type of preservation can yield different types of information, so it can be difficult to be sure that one is dealing with the remains of the same original taxon. A petrifaction can provide abundant information about the detailed cell structure of the plant organ, but it can be difficult to interpret what it looked like as a whole. Similarly, an adpression can show clearly the shape of the organ, but may yield little of its structure.

The main drawback of the form-taxon nomenclature is that it significantly inflates the species list for a given assemblage and can give a misleading impression of the extent of the original biodiversity. However, the processes of transport and fossilization that the plant fragments have been subjected to before preservation have already distorted this information so dramatically (see above) that the nomenclature issue is not really a major problem in this context. The greater accuracy that the form-taxon method provides to the palaeobotanist for recording the plant fossil record far outweighs the problem of inflated biodiversity.

Another problem sometimes encountered with the form-taxon approach is 'what do you call the whole plants?' It does not normally arise with small, herbaceous plants which can be easily reconstructed and are normally named in a similar way to living plants. It is with the larger tree-sized plants that there are difficulties. Even here, however, it is mainly a theoretical problem, because there are very few accurately reconstructed tree-sized plants for which a whole-plant species name would be needed. Many of the reconstructions seen in textbooks have been synthesized from various pieces of information. It may be known that a form-species of cone and a form-species of stem were in organic connection, and that a very similar form-species of stem was attached to a form-species of rooting structure. However, we have very few examples where we can unequivocally reconstruct all of the parts of what was originally an organic, whole-plant species. The reconstructions provide us with a model of what the general groups of plants probably looked like, but they are not images of actual plants that merit formal botanical nomenclature. Hence, it is adequate to give these theoretical reconstructions informal names, such as the '*Lepidodendron* tree' in Text-figure 13, usually derived from the formal name of the most well-known organ (in this case the stem).

WHY DO WE STUDY PLANT FOSSILS?

There is evidence that man had come into contact with the fossilized remains of plants even in prehistoric times; for example, a piece of fossilized wood was found in a Scottish neolithic hearth, which had presumably been placed there by an unsuspecting prehistoric cook! The serious study of plant fossils, however, started in the eighteenth century, and developed as a major discipline in the nineteenth century (for a more detailed discussion of this subject, see Chapter 10). Why has it attracted so much attention from both geologists and botanists?

The fossil record, even though incomplete and biased, is the only direct means of finding out what plants were growing in the past. Most other sources are based around extrapolations into the past from present-day data, which require many assumptions in the analysis. At least the fossil record can show us many of the extinct plants that lived in the past and, even with its limitations, it still has much to tell us of the evolutionary history of plants.

There are other messages that the record can provide. Palaeofloristic work (looking at the geographical distribution of plant fossils) can help in palaeogeographical reconstructions. The apparent anomalies in the distribution of Carboniferous plant fossils relative to the distribution of the continents today was one of the arguments used by Wegener in developing his continental drift model in the 1920s. Wegener's hypothesis formed the basis for the plate tectonics model now used to explain many large-scale processes in the earth sciences.

Changes in the distribution of floras through time can also be an extremely valuable proxy for past climate change. So can changes in leaf shape (leafy physiognomy) especially among the angiosperms. Variations in stomatal density have been used to estimate both long- and short-term fluctuations in atmospheric CO_2. Even the presence of charcoalified plant fossils is now thought to provide an important constraint on estimates of atmospheric O_2; there cannot have been wildfire (which is thought to generate most charcoal) if the atmospheric O_2 was below a certain level.

Plant fossils can also be used to estimate the relative ages of the rocks in which they are found – a science known as biostratigraphy. Animal fossils such as ammonites and corals are generally used for such work in marine strata. Animal groups have also been used in non-marine deposits (a good example is the non-marine bivalves in the Carboniferous of western Europe). On the whole, however, plant fossils are of more use in non-marine rocks, especially in the Upper Palaeozoic, where there seems to have been a very rapid turnover of species. The distribution of plant and spore fossils in non-marine and near-shore marine deposits has been particularly important in helping correlate between the two environments.

There can also be a simple economic imperative to studying plant fossils. Their biostratigraphical value has proved useful in the exploration for natural resources such as oil. Plant remains are also the basis of one of the world's most important energy resources – coal. Most coals are the remains of peat generated in swamps and forests, and maximizing their exploitation can depend on understanding how that peat was formed. This in turn depends on understanding the vegetational dynamics of the original forests, which can only be determined by the study of the plant fossils.

Plant fossils are, therefore, an important tool for the botanist trying to understand the evolution of plant-life, the geologist wanting a means of correlating strata and establishing past continental positions, the climatologist who wants to know about past climates and atmospheres, and the mining engineer who needs help in the exploitation of coal reserves. One reason for the continuing fascination of the subject is that it relates directly to so many different fields.

CHAPTER TWO

EARLY LAND PLANTS

The oldest available evidence of land plants dates back to the middle Silurian, about 425 million years ago. They were the first organisms to make any serious impact on the land. Some bacteria and protoctists seem to have been able to obtain a foothold in earlier habitats but they were never able to reach the physical stature of the plants. Animals do not seem to have attempted to come on to land until there were at least some plants already there. Without plants there was neither food nor shelter, and therefore little to draw an animal to explore the terrestrial habitats, at least as a permanent place to live. The appearance of plants was thus one of the key events in the evolution of life, and was a necessary precursor for the development of amphibians, reptiles and eventually mammals (including ourselves).

ALTERNATING GENERATIONS

Reproduction in all land plants is achieved by alternating sexual and asexual generations. The sexual generation, or gametophyte, produces male and female gametes analogous to sperm and eggs in animals. A male gamete fertilises a female gamete to produce an embryo, from which the asexual sporophyte generation grows. When the sporophyte becomes mature it releases spores, which are dispersed and from which eventually grow new gametophytes. This reproductive strategy probably first evolved among the algae, as there are some living algae that have a version of alternating generations, although they do not produce an embryo.

In bryophytes, the gametophyte generation is dominant and forms the leafy organism that we normally associate with mosses and liverworts. The sporophyte is usually just a stalk with a terminal spore-bearing capsule that grows out of the part of the gametophyte that produced the female gamete. The bryophyte sporophyte is totally dependent on the gametophyte for nutrition and can not live independently of it. In contrast, vascular plants have the sporophyte as the dominant vegetative generation that bears the foliage. In today's 'lower' vascular plants (ferns, club mosses and horsetails) the gametophyte is a small, relatively delicate thalloid structure, whose sole function is to produce the gametes. In seed-plants the female gametophyte and in many cases the male gametophyte have become so reduced that they cannot survive outside of a spore.

ADAPTING TO LIFE ON LAND

There were several physical problems that plants had to overcome before they could live on land. Most centred on the problem of water-supply in their new environment. The newly emergent plants had to be able to support themselves in air, to avoid drying out, and to transport water from that part of their body that was in contact with the wet substrate.

Very small plants can support themselves by maintaining a sufficiently high water pressure ('turgor') throughout their stems, but for most plants this is not enough. The problem was solved by part of the plant tissue developing thickenings on the cell walls, which provided the additional rigidity necessary for support without relying on turgor pressure. Some early land plants achieved this through having a strengthened outer part of the stems, known as a sterome. This can still be seen today in some mosses. However, a more successful solution proved to be a rigid central rod along the plant, known as a vascular cylinder or stele. The main strengthened part of the stele consists of tissue called xylem, made up of rigid cells known as tracheids. The walls of the tracheids have distinctive thickenings of lignin, that can be helical, annular or scalariform (ladder-like), or they may be completely covered except for localised holes called pits. Tracheids can sometimes be recognised as isolated fragments in the fossil record, although care has to be taken not to confuse them with parts of certain animal fossils, such as graptolites. A stele allows a plant to develop an upright stature and to out-compete most other non-vascular plants (i.e. plants without a stele, such as bryophytes and algae) for light. A little later, probably in the Middle Devonian, plants developed a process of secondary growth of the stele to produce woody tissue and hence the giants of the plant world – trees.

But the success of the stele does not lie just in its ability to support the plant. It also solved another of the major hurdles that plants had to overcome to be able to live on land: the ability to transport water from one part of the plant to another. As soon as they have grown to their full size, tracheids die, leaving the thickened cell-walls to form an elongate, hollow conduit along which fluids can pass. Neighbouring tracheids are connected to each other via small sieve-like structures or pits, and thus the entire xylem becomes a route along which fluids can move from the base of the plant to the upper parts where photosynthesis is taking place. Another part of the stele involved in transporting fluids around the plant is the phloem, which, unlike xylem, consists of still-living thin-walled cells. Phloem has no strengthening function, but transports the products of photosynthesis around the plant.

The problem of drying out was partly prevented by most of the aerial surfaces becoming covered with a protective layer called a cuticle. This also doubled as a protective layer against attack by pathogens. However, plants cannot be hermetically sealed from the atmosphere, as their physiology requires gases, in particular CO_2 for the all important photosynthesis and O_2 for respiration. Thin leaves can absorb gases directly through the cuticle but larger thicker leaves need to admit air to come into contact with their inner tissues, otherwise there cannot be sufficient gaseous exchange. This is achieved by small pores (stomata) through the outer surface. A pair of cells (guard cells) usually controls the size of the pore. The bulk of water loss (transpiration) is also through these stomata, so the plant must balance gaseous exchange against water loss to survive. Hence, the pores can be opened when the plant needs to exchange gases with the atmosphere and there is sufficient water for transpiration, but closed when it needs to conserve water.

A cuticle cover is mainly found in the sporophyte of extant plants. The gametophyte generation needs moist conditions for fertilisation to occur, and so there would be no ecological advantage in it having a cuticle to prevent drying out. However, the spores produced by the sporophyte, which are the main agent of dispersal for most plants, do need some protection as there may be a long wait between release by the parent and eventual germination. Spores are covered by a protective layer, different in composition but similar in function to cuticle, known as sporopollenin. In the very early evolutionary history of plants, there was some variation in the manner of production of spores, but plants soon settled on production in clusters of four or tetrads. The three-dimensional tetrad configuration of the spores in the early land plants resulted in a distinctive Y-shaped triradiate mark on the surface of each spore. Triradiate marks can still be seen on the spore surface of many living plants, although some groups produced the spores in a linear tetrad, resulting in a single, linear (monolete) mark on each grain.

THE RHYNIE CHERT FLORA

Although the Rhynie fossils are not the oldest known vascular plant remains, they are by far the best preserved and much of what we know about early land plants arose from work on this flora. Early land plant remains have been found in Upper Silurian and Lower Devonian rocks since research began into rocks of this age in the early 1800s. The problem was that, because they were primitive, the fossils tended to be small and show relatively few characters. As the preservation was not always that good, it was difficult to be certain as to whether these fossils were truly primitive land plants, or were merely poorly preserved fragments of plants such as mosses, or even algae. The debate continued through the nineteenth century and with the limited data then available, the problem seemed intractable.

The breakthrough came in the very early years of the twentieth century, when fragments of chert were found in a dry stone wall near the village of Rhynie in central Scotland. The chert contained what appeared to be numerous round structures, initially thought to be vesicles containing minerals produced by volcanic activity. Only when the chert was examined in thin section was it realized that these structures were in fact petrified stems of the plant we now know as *Rhynia gwynnevaughanii* (Pl. 1). There is no natural exposure of the chert, which occurs in a very small area beneath the soil in a field. However, the importance of the fossils

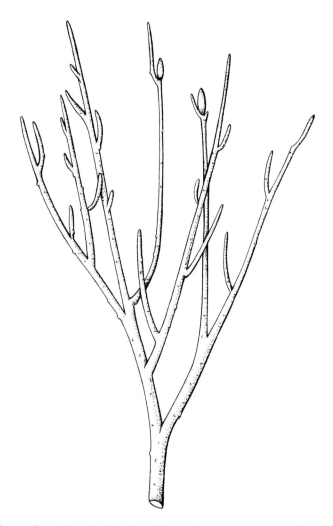

Text-figure 2. The primitive land-plant *Rhynia gwynnevaughanii* Kidston and Lang based on the Early Devonian petrifications of the Rhynie Chert, Scotland (x 1). Drawn by Annette Townsend, based on the work of D. S. Edwards.

was immediately recognised and a series of trenches were dug to obtain additional material and to see how they were positioned in the rock. The result was a classic series of papers by Robert Kidston and William Lang, published in the 1920s in the *Transactions of the Royal Society of Edinburgh*. These remain some of the most important contributions to palaeobotany that have ever been published. Since that time, new methods of sectioning the chert have refined our understanding of the Rhynie plants; we now know almost as much about their morphology and anatomy as could be determined from a living plant.

Rhynie Chert was formed in an area of volcanic activity, and represents an almost entire ecosystem preserved *in situ*, including plants, fungi and animals. The area was periodically flooded by water rich in minerals, produced by the volcanic activity, and these fluids preserved the plants and animals in exquisite detail. The Rhynie plants are unique to this locality, even at the generic level. It is nevertheless possible to relate the evidence obtained from Rhynie to the more abundant but less well preserved adpression floras from the Upper Silurian and Lower Devonian, allowing us to make more botanically meaningful inferences about the affinities and evolution of these early plants.

Rhynia gwynnevaughanii is the commonest species at Rhynie (Text-fig. 2), and is probably the most important, providing the basis for the concept of the Rhyniophytina – the most primitive known division of vascular plants. The plant was relatively small, just a few centimetres high. Its basal parts consisted of a mass of creeping rhizomes from which arose vertical, slender stems. The stems showed both dichotomous and lateral branching, the latter being commonest. There were no leaves or spines on the stems, but there were small bulges along their length. The anatomy of the stems, as seen in thin section, was exceedingly simple, with a slender central stele, surrounded by a two-layered cortex and a thin epidermis with stomata (Pl. 1). Sporangia were born often at the ends of the stems, although a lateral branch sometimes grew from just below a sporangium giving the latter the impression of having been laterally attached. As with earlier rhyniophytes such as *Cooksonia*, the sporangia were simple, with no specialised structures to facilitate the release of the spores.

Another vascular plant with a very simple stele found at Rhynie is *Horneophyton lignieri*. It differs from *Rhynia* in two main features. Firstly, the basal part of the plant consisted of small corm-like rhizomes. The sporangia, although borne terminally on the stems, were also rather different. They were in effect little more than cavities in the end of the stem, into which the stele extended to form a columella-like structure similar to that found in many mosses. Also unlike *Rhynia*, *Horneophyton* has a clear dehiscence structure in the apical part of its sporangia. The affinities of *Horneophyton* have never been properly worked out. It is obviously a primitive vascular plant but some have argued that it shows several features connecting it with the bryophytes (e.g. the sporangia). Today, many palaeobotanists avoid the problem by placing it in a class of its own (Horneophytopsida) within the Rhyniophytina.

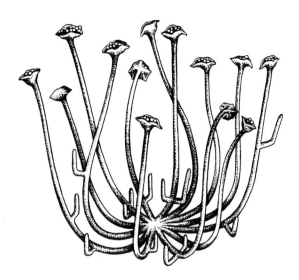

Text-figure 3. The gametophyte *Kidstonophyton discoides* Remy and Hass from the Lower Devonian Rhynie Chert (x 1). It is thought that it is most likely to be the gametophyte equivalent of *Nothia aphylla* El-Saadaway and Lacey. Drawn by Annette Townsend, based on the work of W. Remy and H. Hass.

The nature of the gametophyte in early vascular plants was for many years a problem. Was it a small, ephemeral structure such as found in most living ferns, or a more substantial structure? The answer came with some discoveries by the German palaeobotanist Winfred Remy and his colleagues of several gametophytes in the Rhynie Chert. These were of a similar size to the sporophytes, with the sexual organs borne on vascularised stems, and clearly quite unlike those of any living gametophytes (Text-fig. 3; Pl. 2). So far, gametophyte equivalents have been found for three Rhynie species, including *Horneophyton*, although not so far for *Rhynia*. This important discovery provides the basis for deriving both the more advanced spore-bearing vascular plants (e.g. ferns) in which the sporophyte generation is dominant, and the bryophytes in which the gametophyte generation is dominant.

THE FIRST VASCULAR PLANTS

It is difficult to imagine a more primitive-looking vascular plant than *Cooksonia*. It was only a few millimetres high and consisted of slender, forked stems without spines or leaves, but often with a small spore-bearing body at their end (Text-fig. 4). *Cooksonia* was first recognised in the Silurian and lowermost Devonian of

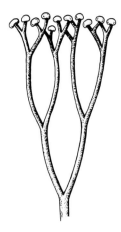

Text-figure 4. The Late Silurian rhyniophytoid *Cooksonia pertonii* Lang (x 2). Drawn by Annette Townsend, based on the work of D. Edwards.

Wales and the Welsh Borderlands (Pl. 3), but have since been found in similar aged strata of Ireland, the Czech Republic, Libya, Kazakhstan and Bolivia. The oldest known examples to date are from the Wenlock Series of Tipperary, Ireland.

The *Cooksonia* fossils are so simple-looking that for many years there was uncertainty as to whether they were truly vascular plants. In the 1930s, William Lang reported slender stems with vascular tissue associated with *Cooksonia*-type sporangia, but no vascularised stems were found with the sporangia attached. In the mid-1980s, Dianne Edwards (of the University of Wales, Cardiff) and her colleagues reported stomata on fertile *Cooksonia* specimens, which strengthened its claim to being a vascular plant. Eventually in 1992, Edwards reported the presence of vascular tissue in unequivocal specimens of *Cooksonia* from the Devonian. It remains to be proved, however, whether the Silurian examples also had vascular tissue.

The debate may on the face of it seem somewhat academic. However, the development of vascular tissue was one of the key factors that enabled the terrestrialisation of life and it is therefore important to understand how and when it developed. There is the added complication that Edwards and her colleagues have discovered that specimens of *Cooksonia* with indistinguishable sporangia can yield markedly different spores and thus presumably belong to different species. It is thus unwise to make the assumption that, as one Early Devonian *Cooksonia* was vascular, middle Silurian *Cooksonia* was also vascular.

We know very little about the appearance of *Cooksonia*, other than it had naked, dichotomous stems with terminal sporangia (Text-fig. 4). These sporangia seem to have been simple, spore-bearing sacs with no specialised structure that facilitated their opening-up to release the spores. We presume that the sporangial bearing stems were borne on some sort of basal structure, either a thalloid-like body or a creeping network of horizontal stems. However, there is no direct evidence of this and many aspects of this earliest of vascular plants remains a mystery.

Cooksonia is now classified among the Rhyniophyta. Before the presence of vascular tissue had been confirmed, it was impossible to be certain that *Cooksonia* was a true rhyniophyte; there was always the possibility that it was a primitive precursor of the rhyniophytes. Because of this uncertainty, Dianne Edwards and David Edwards proposed the term rhyniophytoid, for those very early land plants, which looked like rhyniophytes, but in which vascular tissue had not been confirmed. When this proposal was made in 1986, *Cooksonia* was one of the genera included in the rhyniophytoids. Today, of course, it can confidently be placed in the Rhyniophyta proper.

An example of a rhyniophytoid plant is *Steganotheca* (Pl. 4) from the Upper Silurian of Wales. It has many features in common with *Cooksonia*, including slender, forking stems with terminal sporangia but no spines or leaves. It is a little more robust than *Cooksonia*, the largest specimen being 45 mm long, but the most significant difference is the shape of the sporangia, which are more elongate and have a truncated apex.

True rhyniophytes became extinct at the end of the Early Devonian, although rhyniophytoid-like plants occur through into the Late Devonian. The significance of rhyniophytes lies in that they are probably at the base of the vascular plant evolutionary tree and are therefore ancestors of all the higher plants that we see today, including the gymnosperms and angiosperms.

EARLY NON-VASCULAR PLANTS

At the same time as the vascular plants first appeared, a number of other types of plant were also developing adaptive features to life on the land. The example chosen to illustrate this general group is *Parka decipiens* (Pl. 5). It is a disc-like body, up to 30 mm in diameter, which consisted of two or three layers of cells. On the upper surface of the disc are numerous, small (3 mm in diameter) pillbox-shaped structures containing spores. *Parka* occurs widely in the Lower Devonian of northern Europe. Significantly, it has been found in strata with mud-cracks, suggesting that the original sediment had dried out. This has been used to argue that *Parka* was a plant growing in shallow pools that were periodically subject to drying-out. When the water dried out, the spores were released so that they could be blown to another pool to germinate and grow into another individual plant.

Parka has some similarities to bryophytes and certain green algae. Comparisons have been made, for instance, with a living charophycean alga, *Coleochaete*, which has a very similar thallus and produces oospores with sporopollenin in the wall. It is quite feasible to envisage *Parka* as an evolutionary intermediate between the algae and bryophytes. Whatever its phylogenetic position, however, *Parka* was clearly a non-vascular plant that was trying to get to grips with life on dry land.

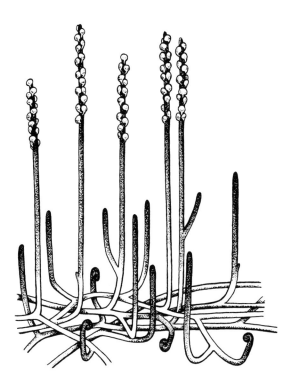

Text-figure 5. The Early Devonian *Zosterophyllum*, showing stems without any spines or other emergences, and terminal clusters of sporangia (x 0.3). Drawn by Annette Townsend, based on the work of P. G. Gensel and H. N. Andrews.

ZOSTEROPHYLLS

Very shortly after the appearance of *Cooksonia*, another type of primitive vascular plant appears in the fossil record: the zosterophylls. Early zosterophylls such as *Zosterophyllum* have some features in common with *Cooksonia*, having naked, dichotomous stems arising from a basal mass of prostrate stems (Text-fig. 5). Unlike *Cooksonia*, however, there is good evidence of the basal part of the plant (Pl. 8). The zosterophylls were generally much more robust plants, probably anything up to 0.5 m high. More significantly, the sporangia were not borne at the ends of the stems, but were laterally attached. The sporangia were also somewhat more sophisticated than those of *Cooksonia*, consisting of two 'valves' that split apart along a definite dehiscence line to release the spores (Pl. 7).

Zosterophyllum was first interpreted as a semi-aquatic plant, with just the terminal parts of the stems with the sporangia protruding above the water. This now seems unlikely, since the stems were covered by stomata and it is most likely that they were fully terrestrial plants. The stomata are of interest because they appear to have just a single, annular guard cell, unlike all other vascular plants where there is a pair of guard cells.

Not all zosterophylls were leafless. The best known example of a leafy zosterophyll is *Sawdonia*, for instance from the middle Lower Devonian (Text-fig. 6). For many years *Sawdonia* stems were referred to *Psilophyton*, a trimerophyte with very similar leafy stems, which will be discussed later. However, it was discovered that some of these leafy stems bore bivalved sporangia very similar to *Zosterophyllum* and clearly belong to the same group of plants (usually referred to the class Zosterophyllopsida).

Another variation on the zosterophyll theme is *Gosslingia breconensis* from the Lower Devonian of Wales (Pls 9–10). This has leafless stems, rather like *Zosterophyllum*, and the individual sporangia are quite similar. However, instead of the sporangia being clustered near the ends of stems, the sporangia are distributed along the length of the stem.

The zosterophylls were important elements of the Early Devonian vegetation, but rapidly declined and became extinct during the Late Devonian. They nevertheless bequeathed a major component of subsequent vegetation as they were ancestral to the club mosses, a group that dominated many Carboniferous and Permian habitats (see Chapter 3).

TRIMEROPHYTES

This was the second major group of plants that arose from the rhyniophytes in the Early Devonian. They retained a number of features from their ancestors among the rhyniophytes, such as stems that tend to be naked or only covered by small spines, and sporangia that are attached to the ends of stems. However, the trimerophytes were morphologically more complex and larger plants than most rhyniophytes, and clearly represent a major evolutionary advance.

The largest trimerophyte found to date is *Psilophyton forbesii* from the Lower Devonian of North America, and was at least 0.6 m high. Where internal anatomy has been demonstrated, such as in *Psilophyton dawsonii* from the Lower Devonian of North America, the stele represents up to about a quarter of the volume of the stem (a reconstruction of this plant is shown in Text-fig. 7). This is much greater than occurred in rhyniophyte stems and may reflect the additional mechanical requirements of these larger plants. The stems of the trimerophyte produced lateral (monopodial) branches. These lateral branches, which show a range of different types of branching pattern, were either sterile or fertile. It has been argued that the clusters of sterile axes, if they became flattened in a plane, might have represented the early origins of complex leaves such as fern fronds.

The fertile branches produced clusters of sporangia at the ends of slender axes (Pl. 11). Unlike the rhyniophytes, the trimerophyte sporangia had a clearly developed dehiscence slit along their length to release the spores on maturity.

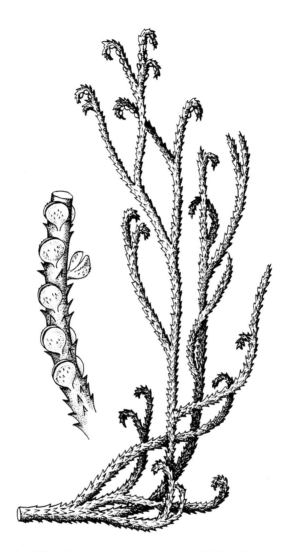

Text-figure 6. The Early Devonian zosterophyll *Sawdonia*, showing the stems with dense spines (x 0.5). Also showing a close-up of part of a fertile shoot, in which the bivalved sporangia are shown, one of the characteristics of the zosterophylls (x 2) . Drawn by Annette Townsend, based on the work of P. G. Gensel.

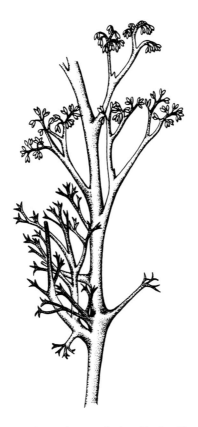

Text-figure 7. Part of a shoot of the Early Devonian trimerophyte *Psilophyton*, showing the characteristic clusters of sporangia (x 0.75). Drawn by Annette Townsend, based on the work of H. P. Banks and his colleagues.

The clusters of sporangia are very similar in most trimerophytes, and can be difficult to assign to a natural species unless they are attached to their parent plants. For this reason, we normally assign such isolated trimerophyte sporangial clusters to their own form-genus, *Dawsonites*.

The importance of the trimerophytes lies in their probable ancestral relationship to at least two other major groups of plant: the ferns and the progymnosperms. As the progymnosperms (which we will return to again later) was probably the ancestral group to all later seed-plants, we have in the trimerophytes the ancestors of most of the higher plants living today.

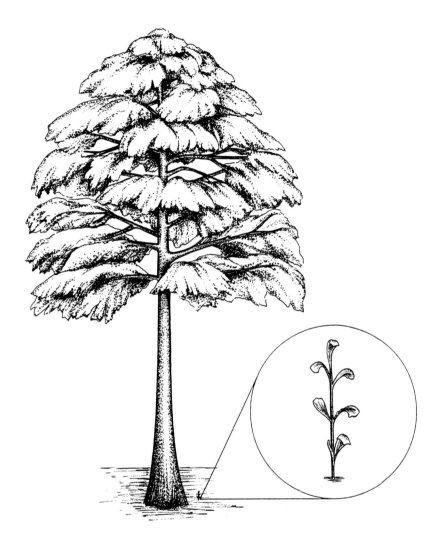

Text-figure 8. One of the earliest types of tree, *Archaeopteris*, showing both a fully mature tree about 25 metres high, and a juvenile plant. Drawn by Annette Townsend, based on the work of C. B. Beck.

PROGYMNOSPERMS

By the Middle Devonian, many of the primitive groups of land plant were becoming quite substantial organisms. However, the basic construction of the stems of these plants limited the size that they could achieve. Plants overcame this limitation by developing secondary growth, in particular secondary xylem (or wood). The primary xylem is produced by a plant at the tips of its stems and roots, where cells form what is called an apical meristem that continuously divides to produce new tissue. A short way behind this growing tip, the new cells differentiate into the main types of tissue in the plant, such as primary xylem and phloem. Secondary growth is achieved by another zone of dividing cells, the lateral meristem or vascular cambium, located between the primary xylem and primary phloem. The resulting secondary xylem, and to a lesser extent secondary phloem, allows the plant to significantly increase the girth of its stems with mechanically strong tissue. The result was the development of trees, some of which have grown to be the largest living organisms to appear on earth.

Secondary wood was for many years recognised in the Middle and Upper Devonian rocks (Pl. 15), without there being any information about the types of plant that would have produced such 'advanced' tissue. It was speculated that perhaps conifers were already in existence at this time, but no coniferous foliage or reproductive structures could be found. The issue was not properly resolved until the 1960s, when the American palaeobotanist Charles Beck demonstrated that some of these Devonian trees bore foliage and spore-bodies (similar to that in Pl. 14) that had hitherto been interpreted as having affinities with the ferns. From this arose the concept of the progymnosperms, for plants that had secondary wood like the seed plants, but which reproduced by spores. A reconstruction of one of these plants, both as a mature tree and a juvenile, is shown in Text-figure 8.

Several different types of progymnosperm are now recognised. The earliest date back to the Givetian (Middle Devonian), such as *Protopteridium* from northern Scotland (Pl. 12). This was probably quite a substantial plant, as the trunk had secondary wood. It did not have distinctive leaves, but rather irregularly branching, three-dimensional clusters of axes, somewhat reminiscent of the foliage of their rhyniophyte ancestors. Small, fusiform sporangia were borne in clusters at the ends of axes, which themselves were laterally attached to a lower-order branch.

Slightly younger are remains of a plant known as *Svalbardia* that have been described from Spitsbergen and Fair Isle (Pl. 13). They show the first signs of proper, planated leaves, although they were deeply divided leaves with very little lamina, and thus somewhat similar to the vegetative branches of certain trimerophytes. The sporangia are also similar to those of *Protopteridium*, being elongate with a longitudinal dehiscence slit, and borne in two rows along the side of a specialised fertile branch or sporophyll.

Fossil leaves of a somewhat more advanced looking progymnosperm *Archaeopteris* occur widely in the Upper Devonian of Europe and North America (Text-

Text-figure 9. Part of a shoot of the progymnosperm *Archaeopteris*, showing both the foliage and sporangia (x 1). Drawn by Annette Townsend, based on the work of H. N. Andrews.

fig. 9; Pl. 14). The sporophyll is comparable with those of the homosporous *Svalbardia*, but at least some species have been shown to be heterosporous. Heterospory is where a plant produced two size-ranges of spore: large megaspores, which germinate to produce a female gametophyte, and much smaller microspores, which germinate to form a male gametophyte. This is widely accepted as a precursor to the appearance of the seed-habits, where the female spores have enlarged to the extent that only one reaches maturity in the sporangium (see Chapter 5). *Archaeopteris* also differs from *Svalbardia* in that it had planated leaves with entire lamina between the veins. There have been disagreements as to the significance of the difference between these two plants, and some palaeobotanists argue that they should be assigned to the same genus. Whatever the outcome, however, it is clear that these plants almost certainly represent the early phases in the evolution of seed-plants from their ancestors among the trimerophytes.

CHAPTER THREE

CLUB-MOSSES

Lycophytes today are a relatively modest group of small plants, many of which have a superficially moss-like appearance (e.g. *Huperzia*), hence their informal name, the club-mosses. However, this gives a totally misleading impression of a group of plants that has played an important role in the history of land vegetation. The lycophytes have the longest fossil record of any group of plants, extending back at least to the Early Devonian, and for part of this time they were dominant elements in the terrestrial vegetation. Especially in the Late Carboniferous, they were the largest known living organisms and formed dense forests over much of the tropical belt.

Modern lycophytes have a relatively simple morphology usually with undivided simple leaves (called microphylls). They reproduce vegetatively, or by spores produced in sporangia that are usually associated with a microphyll. The sporangia may be arranged in fertile zones on vegetative stems, or may be in terminal cones where the modified microphylls are referred to as sporophylls. The spores may be all similar (homospores) or of two kinds (megaspores and microspores).

THE EARLIEST LYCOPHYTES

Baragwanathia (Pl. 16) is a fleshy herbaceous vascular plant that can be closely compared with simple lycophyte species of the large extant genus *Lycopodium*. Its stem branched dichotomously to psuedo-monopodially and was probably recumbent with well-developed adventitious roots that arose directly from the leafy stems. The stem was fleshy and without the outer layer of thickened, fibre-like cells present in many other lycophytes. This interpretation is based on the fact that all that remains in the compression fossils of the stems is the cuticle, the vascular strand and the leaf traces. The vascular strand was only about one tenth of the diameter of the stem so the plant must have supported itself by turgor pressure, as in some of the early rhyniophytes (see Chapter 2). The leaves were simple, about 30 mm long, with entire margins, and arranged in helices. Regions of the plants were fertile with sporangia borne in the axils of unmodified leaves. The sporangia released their small, trilete spores through a slit orientated transversely to the leaf.

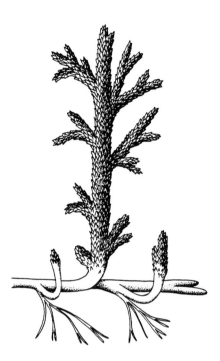

Text-figure 10. The primitive lycophyte *Asteroxylon mackei* Kidston and Lang based on the Early Devonian petrifactions of the Rhynie Chert, Scotland (x 0.5). Drawn by Annette Townsend, from the work of D. A. Eggert.

Specimens bearing sporangia are, however, uncommon which suggests that the plants may have also reproduced vegetatively by simple fragmentation.

Baragwanathia longifolia is the oldest known lycophyte species and has been found at two localities in Victoria, Australia. However, its exact age has been the subject of much debate for a number of years. The earliest occurrence has been interpreted as some 424 million years old (Gorstian, Lower Silurian) based upon the identification of associated graptolites as *Monograptus uncinatus*, a species restricted to the Gorstian. If this were correct, *Baragwanathia* would be only marginally younger than the oldest known examples of *Cooksonia* (Homerian in age), which are generally accepted to be the first land plants. However, there is a doubt about the determination of the graptolite species, since some of its characters appear to range into those of other, Devonian graptolite species. The floral assemblage at the 'Gorstian' locality is also conspecific with that found at a later Pragian (middle Lower Devonian) locality. It seems unlikely that there was no significant evolutionary change in the floral assemblage during the 24 million years separating the two supposed dates. The earlier date is, therefore, very suspect and a Pragian age for both assemblages seems more realistic. This later date also fits in with the accepted ideas of the general theory of evolution in the plant kingdom.

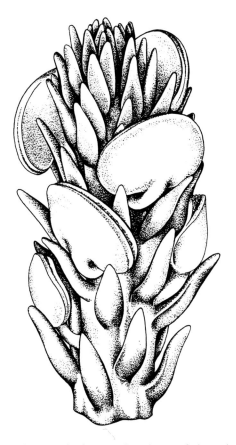

Text-figure 11. The terminal part of a shoot of the primitive lycophyte *Asteroxylon mackei* Kidston and Lang based on the Early Devonian petrifactions of the Rhynie Chert, Scotland (x 4). Drawn by Annette Townsend, from the work of W. G. Chaloner.

LYCOPHYTES OF THE RHYNIE CHERT

Asteroxylon mackei (Text-fig. 10) is another plant that Kidston and Lang described from the famous Lower Devonian Rhynie Chert flora in Scotland (see Chapter 2). It was superficially like the living lycophyte *Huperzia selago* in appearance, with a prostrate, creeping rhizome that produced erect stems clothed with helically arranged microphyllous leaves. Its sporangia were kidney shaped and appear to have arisen directly from the stem (Text-fig. 11). They were relatively large and projected beyond the leaves, which has led some palaeobotanists to suggest that they were actually attached to the upper surface of a leaf. The sporangia dehisced apically to release its one kind of trilete spore. The plants were, therefore, homosporous.

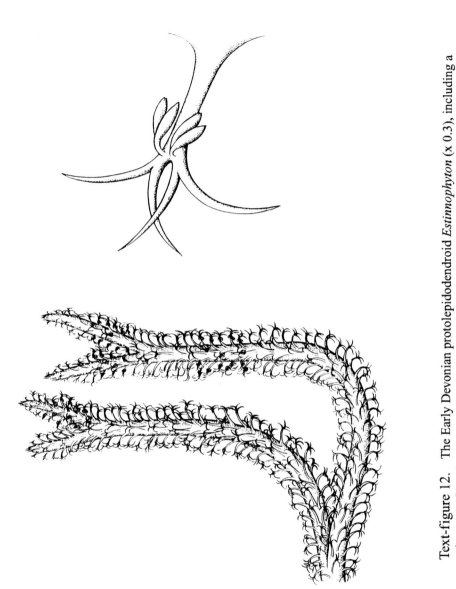

Text-figure 12. The Early Devonian protolepidodendroid *Estinnophyton* (x 0.3), including a close-up of its characteristic divided leaves (x 6). Drawn by Annette Townsend, based on the work of M. Fairon-Demaret.

Because *Asteroxylon* comes from the Rhynie Chert, its anatomy is quite well known. It has a central vascular strand with a stellate, mesarch xylem (Pl. 17). The stem apex is flat with several meristematic initials and, as such, resembles the apex of living *Lycopodium reflexum* Lamarck. Leaf traces run from the central vascular bundle to the base of the leaves but do not actually enter them.

There has been some debate about the exact botanical affinity of *Asteroxylon*, with some experts believing it to be a lycophyte while others prefer to think of it as a lycophyte-precursor. In our view, however, the close comparisons of overall morphology, and stellar and apical anatomy seem to outweigh the differences of potential sporangial attachment (about which there is anyway some doubt) and the lack of vascular supply to the leaves. *Asteroxylon* is surely a lycophyte.

PROTOLEPIDODENDRALES

Other Devonian lycophytes can be distinguished as a group, the Protolepidodendrales, by the presence of leaves that branch at their tips and are attached by swollen bases to the stem. *Estinnophyton* (Text-fig. 12) is the most completely known genus because its compressed parts have been found very well preserved. It had creeping dichotomizing stems that occasionally turn upward to bear fertile zones, rather like *Huperzia selago*. Its stem anatomy is simple, although more fragmentary than that of *Asteroxylon*. *Estinnophyton* has its leaves (microphylls) divided apically typically into five, but sometimes up to seven, parts. They have a little flap of tissue, a ligule, near the base of their upper surfaces. Sporangia are borne singly on the upper surfaces of unmodified microphylls. They are stalked and attached at some distance from the base. Only one type of spore has been recovered from the sporangia, so *Estinnophyton* appears to have been homosporous. These plants are therefore very unusual because all other known ligulate lycophytes (e.g. the living *Selaginella*) are heterosporous.

CUTICLES AND PAPER COAL

The outermost epidermal layer of cells of the aerial parts of most vascular plants is covered with a complex polymer of hydroxy-acids and other substances, called cutin. This 'skin' of cutin is called cuticle, and provides one of the structures used by plants for reducing water loss. To be effective, cuticle has to cover the surface and extend down into the walls between the epidermal cells. In some plants the cuticle extends around the epidermal cells and into the walls of a specialized subepidermal layer called the hypodermis.

The physical and chemical 'toughness' of cuticle means that it can withstand the destructive processes involved in fossilization. A technique called maceration, involving the use of an oxidizing acid followed by an alkali, separates the cuticle

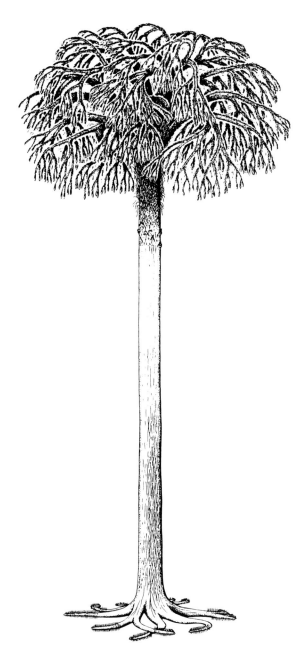

Text-figure 13. The *Lepidodendron*-tree, which was 30–40 metres high, was one of the arborescent lycophytes that dominated the tropical swamp forests of the Late Carboniferous. Drawn by Annette Townsend, based on the work of B. A. Thomas.

from the rest of the coalified adpression and makes it transparent and ready for microscopic study. Such cuticle preparations show the outlines of the epidermal cells including stomatal guard cells, hairs and trichomes. Epidermal details, revealed by such cuticular studies, can be vital in identifying and comparing plant fossils (see Chapter 1 for a further discussion on cuticles).

Cuticular studies on fossil lycophytes have revealed many interesting facts. Epidermal details of leaf cushions help distinguish species, stomatal guard cells are now known to be sunken in pits and ligule pits shown to be cuticle covered.

Sometimes the preservation process has been relatively slow, resulting in extensive decay of the plant tissue. This process has left layers of plant fossil material that are little more than cuticle, which are referred to as Paper Coals. The two best known examples of such preservation are the Jurassic Paper Coal of Roseberry Topping in Yorkshire, England, consisting mostly of *Pachypteris* gymnosperm leaves (see Chapter 7), and the Russian Lower Carboniferous Paper coal of the Moscow Basin, which consists entirely of lycophytes (Pl. 19). These lycophytes were first described in 1860 as *Lepidodendron*, but have since been referred to the lycophyte genus *Eskdalia*. The slabs of cuticle are from lycophyte stems and the oval holes mark the attachment points of leaves. Epidermal cells of the stem cuticle are clearly shown, but there are no stomatal guard cells. The small sock-like structures hanging down from the upper angles of the holes are the cutinized inner walls of ligule pits. Fragments of thinner cuticle found attached to the edges of the holes are portions of cuticle from the basal portions of the leaves, and on them can be seen epidermal cells and pairs of stomatal guard cells.

ARBORESCENT LYCOPHYTES

Some Carboniferous lycophytes grew to 50 m in height with a crown of branches, these being the 'trees' of the coal-forming swamps (Text-fig. 13; Pls 26–27, 30). Although they resembled trees in appearance, they had rather different growth patterns. Sporelings grew upward and rapidly expanded in stem width until they resembled small poles covered in leaves. These stems increased in height and girth until they were about 30–40 m tall, then they divided apically many times. Each successive division produced two smaller branches until the smallest terminal shoots were no more than a few millimetres thick. Unlike true trees, the size of a lycophyte branch is no indication of relative age.

This unusual and rapid growth pattern was only possible because there was a support system of thickened cells in the outer parts of the stem (Pl. 18) rather than increasing central wood as in true trees. The growth pattern, especially the expansion in girth, produced in a series of leaf, 'bark', shoot and fructification abscissions that resulted in these parts becoming fossilized as isolated organs. Each type of fossil organ has to be treated separately for taxonomic purposes because the number of organs that have been found connected is very few. Therefore, there may be several fossil genera of each type of organ that each can have a large number of species within them.

EARLY ARBORESCENT LYCOPHYTES

Cyclostigma is the name given to fossil remains of an early arborescent lycophyte found in the Upper Devonian of the British Isles, China and Japan (Pl. 20). It had a main trunk up to about 8 m tall and 0.3 m in diameter with a crown of dichotomizing branches. Its leaves were variable in size but often attained 150 mm in length on the smaller branches. Leaf fall left circular or oval scars that each show a small central vascular scar. Neither ligules nor ligule pits have been observed. Some of the larger branches show a mesh of fine anastomosing lines indicating an expansion of the outer tissues of the stem.

The terminal cones were about 60 mm long and 30 mm in diameter with sporophylls. A radially elongated sporangium was attached to the upper surface of the basal portion of each sporophyll. The sporophyll laminae were about 160 mm long and very similar in appearance to the leaves.

Lycophyte stems may be found with leaves still attached or with the scars showing where the leaves were once attached. The leaves in many instances are attached by swollen bases, which remain when the linear portions of the leaves are shed. These swollen leaf bases are called leaf cushions and carried on the process of photosynthesis through their many stomata.

LEPIDODENDRON AND THE COAL MEASURES SWAMPS

Lepidodendron is the generic name that is given to a common Carboniferous lycophyte stem that has longitudinally elongated leaf cushions (Pl. 29). Where the leaves have been shed, leaf scars are left in the centre of the cushions (Pl. 21). Each shows three scars. The central scar shows the vascular bundle, the two lateral scars the positions (parichnos) of aerating canals that ran into the leaves. Lower down on the cushion surface there are sometimes two more parichnos prints. Such aerating canals most probably facilitated gaseous exchange associated with photosynthesis and recycling of oxygen and carbon dioxide. Above the leaf scar and sometimes adjacent to the upper angle, is another small print marking the entrance to a small pit in the bottom of which was attached the ligule.

The terminal shoots of arborescent lycophytes are notoriously difficult to distinguish and they are usually identified as one of only four species. This has to be wrong considering the much larger number of species of stems with recognizable leaf cushions. Future studies involving epidermal studies may help resolve this problem.

While *Lepidodendron* is one of the best known of the Carboniferous arborescent lycophytes, there were other genera represented in the Coal Measures forests. *Sigillaria* is the name given to a type of bark in which the leaf cushions are arranged in verical rows (Pl. 31). It was probably smaller than the *Lepidodendron* plant, with a trunk that did not branch when mature, and which probably favoured the marginal, slightly drier parts of the swamps. *Lepidophloios* with its distinctive horizontally elongated leaf cushions (Pls 24–25), seem in contrast to have favoured the wetter parts of the forest.

Another way to distinguish between different types of arborescent lycophyte is to relate the leafy shoots to the species of cone that they bear (Pls 32–33). They reproduced by forming spores in specialized cones. All the cones had the same basic morphology of helically arranged sporophylls each with a sporangium attached to its upper surface. Genera of cones are distinguished by a combination of sporophyll shape and sporangial spore content. Almost all arborescent lycophytes were heterosporous thereby producing both megaspores and microspores. Within the heterosporous lycophytes, some cones had both kinds of spores and are said to be monoecious, while others had one kind and are said to be dioecious.

Lepidodendron bore cones on the ends of its terminal shoots. They belong to the monoecious genus *Flemingites* and varied in size from about 10 to 100 cm in length. Such cones have been found permineralized in coal balls giving us details of their anatomy, and as adpressions allowing us to prepare their spores and make direct comparisons with genera and species of dispersed spores.

Flemingites cones have helically arranged sporophylls that are attached at approximately right angles to the cone axis. This 'horizontal' part of the sporophyll is called the pedicel and has an elongate sporangium attached to its upper (adaxial) surface. A ligule in a pit is situated distally to the sporangium. Beyond the sporangium and the ligule the sporophyll turns abruptly upwards as the leaf-like lamina. Sometimes there is a downward projecting small 'heel'. Individual sporangia produce only one kind of spore and usually it is the more apical ones that are microsporangiate. The megaspores are usually quite distinct and they can be used to help distinguish species of otherwise similar cones. Most are about 1 mm in size, spherical or with a resemblance to an inflated sack with a triangular projection where the three contact faces are. The body of the spores may be smooth or covered in spines, respectively being closely comparable to the dispersed spore genera *Lagenoisporites* and *Lagenicula* (Pl. 34). The microspores are about 25–35 μm in diameter, trilete, with smooth contact faces and granular distal surfaces. Because there are thousands of microspores in each sporangium, there is some natural variation amongst them. Nevertheless, they all fit within the limits of the dispersed spore species *Lycospora* and with careful observation different populations can be identified in different species of cone.

Some arborescent lycophytes developed a reproductive strategy that had some features in common with seed-plants, the best known of these being the plant that had *Lepidophloios* branches and *Lepidocarpon* cones. In the female cones, each sporangium contains but a single megaspore and the sporangium is partly enclosed by lateral extensions of the sporophyll (Pl. 36). The sporophyll is therefore partly fulfilling the protective role of an integument in a gymnosperm seed (see Chapter 6). Furthermore, dispersal was achieved by the release of the entire sporophyll unit from the cone (Pl. 35) rather than shedding of the individual spores as normally happens in pteridophytic plants. Although there is no suggestion that these aborescent lycophytes were related to gymnosperms, it is interesting that they started to develop a broadly similar reproductive strategy. Why they did not continue to evolve and flourish, as did the gymnosperms, is a mystery.

STIGMARIA

The arboresent lycophytes with their tall axes and large aerial branching crowns clearly needed an extensive rooting system to provide for both the physiological needs of the plants and stability in the soft sediments in which they grew. Such systems, which are physiologically but not morphologically true roots, are given the name *Stigmaria* (Pls 26–29). They spread out, more or less horizontally, from the bases of the main stems by dichotomous branching. As they spread the stigmarian axes increased in girth, partially through secondary xylem production, but mainly by the formation of extra cortical tissues. Each growing stigmarian apex was a rimmed depression terminated by a protective plug of parenchymatous tissue. True roots were formed by the growing apices and radiated in all directions. When fully grown the roots could be up to 0.5 m or more long and the vertical ones may have projected out of the sediment into the overlying water. The roots had a large central canal, which provided a pathway for gases and possibly permitted gaseous exchange with the surrounding waterlogged sediments or the overlying water. As the stigmarian axes expanded through secondary growth, the older roots were shed leaving characteristic circular scars on their surfaces. There are virtually no characters that can be used to distinguish the stigmarian bases of different species of parent plants so practically all of them are placed in the same species *Stigmaria ficoides* (Sternberg) Brongniart.

Many of these stigmarian bases have been preserved as casts in sandstone. After the plants died and the aerial parts of the plants disintegrated is was possible for the rotting stigmarian axes to be infilled with fine sediment. In time this would harden producing a cast. The almost perfect example illustrated here from the Manchester Museum (Pl. 27) is the specimen described by Williamson in his original work on *Stigmaria*. It was discovered by quarrymen working near Bradford in 1886 and removed to Manchester at Williamson's own expense. The diameter of the stem and the spread of its rhizophores are 1.3 m and 9.0 m respectively. The first formed stigmarian axes are 0.5 m in diameter and after two dichotomies the most terminally preserved axes are 50 mm in diameter. As the best known preserved stigmarian apex is 50 mm in diameter, it can be deduced that the Manchester *Stigmaria* is virtually complete.

Many other stigmarias have been found since then in quarry excavations and in eroding sea cliffs but are usually not so well preserved. The two best known sites are the small grove preserved at Victoria Park in Glasgow, Scotland (Pl. 26) and the many horizons of fossil forests at the famous coastal site at Joggins in Canada. From sites such as these, it has been possible to estimate that there were approximately 1500–2000 of these large tree-like lycophytes per hectare in the Late Carboniferous forests, which is sigificantly more than the density of trees found in today's tropical rain forests.

REDUCTION IN SIZE

The warming of the climate in the Late Palaeozoic and Early Mesozoic brought about the extinction of the arborescent and subarborescent lycophytes. All lycophytes surviving into and beyond the mid-Jurassic were herbaceous. Today the survivors are the leafy *Lycopodium* (often divided nowadays into four smaller genera) and *Selaginella* species, the monotypic *Phylloglossum drumondii* and the cormose species of *Isoetes*.

Pleuromeia (Text-fig. 14) is known from many early to middle Triassic localities in the mid-latitudes of the Northern Hemisphere. They ranged in height from 0.2 to 3 m. The vertical stems were unbranched and covered in simple lanceolate leaves that could be up to 100 mm long on the largest stems. Leaves were shed in the basal regions as the stem grew. Eventually a terminal cone was formed of rounded sporangia attached to the upper surface of distinctive broad ovate sporophylls. The species were all heterosporous although individual cones were monosporangiate, containing either microspores or megaspores. The smallest species had simple bulbous rooting bases while the largest had four-lobed bases reminiscent of the lepidodendralean *Stigmaria*.

There are several localities where *Pleuromeia* has been found in stands suggesting that it grew as the dominant vegetation over large areas within coastal plain or deltaic drainage basins. There is still some debate over whether some such stands may have grown in brackish environments such as marine embayments.

An understanding of *Pleuromeia* is central to our interpretation of the evolution of modern *Isoetes*. In 1956, Magdefrau proposed a simple reduction series from the Carboniferous lepidodendraleans through the Triassic *Pleuromeia* and the Mesozoic *Nathorstiana* to living *Isoetes*. Modern theories now suggest that the initial member of the reduction series is more likely to have been the subarborescent lycophyte *Chaloneria* which, in fact, hardly differed from *Pleuromeia*. Documented changes which result in the modern *Isoetes* include the reduction of the axis, the development of sunken sporangia with a covering velum, a change from trilete to monolete microspores, and the development of a swollen ligule base (glossopodium).

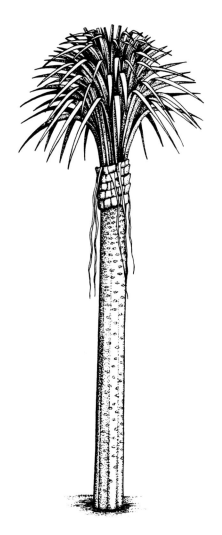

Text-figure 14. The Early Triassic lycophyte *Pleuromeia* was significantly smaller than the Late Carboniferous lycophytes, probably only about 2 m high, but was nevertheless larger than club-moss growing today. Drawn by Annette Townsend, based on the work of G. J. Retallack.

MODERN HERBACEOUS LYCOPHYTES

Although the majority of Carboniferous lycophytes were either bushy or tree sized, there were some that were truly herbaceous. Some are preserved as adpressions and others as permineralizations. Some are fertile while others are merely fragments of vegetative shoots.

Many of the Carboniferous herbaceous lycophytes (e.g. Pl. 37) had a striking resemblance to the extant herbaceous *Selaginella*. Some, such as *Paurodendron fraipontii* were sprawling and sparsely branched with spirally arranged microphyllous leaves and terminal cones. Others had thin stems that bore ranks of different sized leaves and terminal cones thereby appearing very similar in appearance to such extant species as *Selaginella krausiana*. There is a major difference, however, in that the Carboniferous forms have three ranks of leaves while the living species have two. We have no idea as yet when the reduction in the number of ranks occurred.

Isophyllous forms first appear in the Lower Carboniferous of Scotland, and since the species are loosely defined they have very long stratigraphical ranges. The anisophyllous ones made their appearance in the Bolsovian, species-poor intra-montane Saar-Lorraine Basin in Europe rather than in the widespread paralic swamps which were dominated by the arborescent lycophytes. The stimulus for their evolution may have been triggered by different competitive forces which favoured a creeping habit rather than an erect one, although how it occurred may be a question that will never be answered. By the Westphalian D herbaceous forms had spread over an area between Nova Scotia, Canada to Zwickau in eastern Germany. At first sight the anisophyllous forms are all very similar but several species can be distinguished on leaf shape and epidermal characters, which are features used in identifying living species of *Selaginella*.

CHAPTER FOUR

HORSETAILS

Horsetails have never been a particularly diverse group of plants and today there is only one living genus (*Equisetum*) with about twenty species. They nevertheless form a remarkably successful group, now growing throughout the world in a range of habitats, and only absent from Australia, New Zealand (except for some naturalized introductions) and Antarctica.

The horsetails are very distinctive, with branches and leaves borne in whorls on ribbed stems. The stems are mainly hollow, or have a central canal filled only with soft, easily degraded tissue; only at the nodes (the positions along the stems where leaf or branch whorls occur) is there a diaphragm of tissue stretching across the width of the stem. Because their stems are so often hollow, horsetail pith casts are regularly preserved in the fossil record (discussed in more detail in Chapter 1). In all living species and many of those found in the fossil record, the leaves of each whorl are fused basally, to form a distinctive sheaf around the stems. In most Palaeozoic forms, however, the leaves are not fused or are only joined by a very narrow collar of tissue. Reproduction is by spores produced by cones borne terminally on stems. Except in some of the more primitive Palaeozoic forms, the sporangia are borne within the cone on specialized peltate (mushroom-shaped) sporangiophores, which are quite different from the sterile foliage (unlike ferns and many club mosses). The spores produce relatively large gametophytic prothalli, and often have distinctive strap-like appendages called elaters that probably help with their dispersal.

ORIGIN AND SYSTEMATIC POSITION OF THE HORSETAILS

There is unequivocal fossil evidence of horsetails as far back as the Late Devonian. They probably originated from a group of early fern-like plants known as the Cladoxylales, which includes *Calamophyton* (see Chapter 5) and which itself arose from the trimerophytes. *Calamophyton* and its close allies in fact have many features in common with the early horsetails, such as tufts of sterile axes that resemble leaves. Their sporangia were borne on the recurved tips of slender axes and were clustered into loose cones. They were once regarded as being most closely related to the horsetails. Only when the anatomy of the stems was discovered was it realized that their affinities lay closest to the ferns. Nevertheless, it is likely that the cladoxylaleans represent the ancestors of the horsetails.

Horsetails have a very similar mode of reproduction to the club mosses and ferns, and many botanists still group them all together into the division Pteridophyta. However, this type of reproductive cycle is a primitive feature that has been retained from their Palaeozoic common ancestors, and suggests that the fern, club moss and horsetail lineages have been independent at least as far back as the Devonian. It therefore seems more reasonable to assign them to their own divisions or phyla. There has been disagreement as to the formal name of the horsetail division, different authors referring to it as the Equisetophyta, Sphenophyta or Arthrophyta; here we will use Sphenophyta.

Three orders are normally recognized within the horsetails: the Pseudoborniales, the Sphenophyllales and Equisetales. The first two are now extinct. The third includes the family of living horsetails, the Equisetaceae, plus two Palaeozoic families the Archaeocalamitaceae and Calamostachyaceae.

PSEUDOBORNIALES

The oldest undoubted example of a horsetail is found in Upper Devonian strata from around the Arctic Ocean (Bear Island in the North Atlantic, and Alaska). It was the size of a small tree, up to 20 m high, with a trunk about 0.6 m wide at the base. The trunk had horsetail-like nodes along its length, but each node only gave rise to one or at most two branches, each of which could be up to 3 m long. The branches show longitudinal ribs, which like the archaeocalamites (see next section) do not alternate at the nodes. These first order branches bore lateral secondary branches, which themselves then bore the ultimate branches with leaves attached (Pl. 38). These leaves, borne in distinct whorls of four, were much more complex than most horsetail leaves. They forked two or three times near the base, and each segment is then further divided in a pinnate-manner.

Cones were borne at the ends of the first order branches, in the upper part of the plant. They consist of alternating whorls of bracts and sporangiophores (modified branches bearing sporangia), the latter bearing about 60 sporangia at the ends of recurved axes.

Pseudobornia seems to represent an intermediate position between the horsetails and their cladoxylalean ancestors. It had clear horsetail characteristics: the ribbed stems, the whorled arrangement of the leaves and sporangiophores, and sporangia borne in cones. On the other hand, there are features that link it with the cladoxylaleans such as the sporangia being borne on recurved axes and the complexity of the leaves. On the whole, however, most palaeobotanists regard it as a primitive horsetail, placing it in its own monotypic order, the Pseudoborniales.

ARCHAEOCALAMITACEAE

Horsetails first seem to have become common components of land vegetation in the Early Carboniferous. Rocks of this age regularly yield fragments of their stems and to a lesser extent of their leaves and cones. The stems are referred to as *Archaeocalamites radiatus* and, like most horsetails, are characterized by ribs running along the stem, and transverse furrows representing the positions of the nodes. Petrified specimens show that there was a well-developed pith cavity along the stem, the first time this horsetail characteristic is found in the fossil record. Unlike modern horsetails, however, many *Archaeocalamites* stems show evidence of secondary wood, indicating that they could have been parts of large plants.

There are a number of primitive characteristics in *Archaeocalamites* that link it with the Pseudoborniales discussed previously, and their cladoxylalean ancestors. The leaves fork in a complicated way, somewhat resembling enlarged *Pseudobornia* leaves but without the ultimate pinnate segments. Also, the ribs do not alternate as they pass over the nodes, as they do in the later horsetails. The loose cones, usually referred to the form-genus *Pothocites* (Pl. 44), have sporangia borne in groups of four on sporangiophores with a peltate head, similar to those found in calamostachyacean cones. Also like the Calamostachyaceae, they were sometimes heterosporous (Pl. 46). Unlike the Calamsotachyaceae, however, *Pothocites* usually does not have bracts, making it seem more reminiscent of the modern *Equisetum* cones. Furthermore, the microspores do not have the distinctive elaters found in both the Calamostachyaceae and modern horsetails (tiny strap-like extensions from the spore surface, which become springy when dry and help disperse the spore). This apparent mixture of primitive and advanced features has resulted in *Archaeocalamites* and *Pothocites* being assigned to their own family, usually known as the Archaeocalamitaceae.

Archaeocalamites was for long regarded as a characteristic plant of the Early Carboniferous, with no known occurrences in rocks of Late Carboniferous age. It was widely believed that they were out-competed by their more sophisticated descendants the Calamostachyaceae. This view may have to be revised in the light of the recent discovery of Permian fossils that look identical to *Archaeocalamites* stems and leaves. Although the identifications need to be confirmed by the discovery of fertile material, it looks likely that *Archaeocalamites* was only displaced from the wet, lowland habitats and that they were perfectly well adapted to survive in the drier, extra-basinal areas. This may have consequences for our understanding of the origin of the Equisetaceae, which will be discussed later.

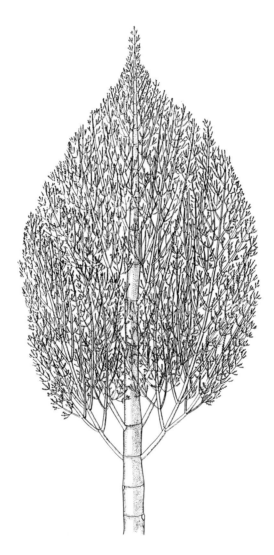

Text-figure 15. The Late Carboniferous sphenophyte *Calamites*. They were trees up to some 20 m high that mainly surrounded the margins of lakes within the tropical forests. Drawn by Pauline Dean, based on the work of C. F. Grand'Eury.

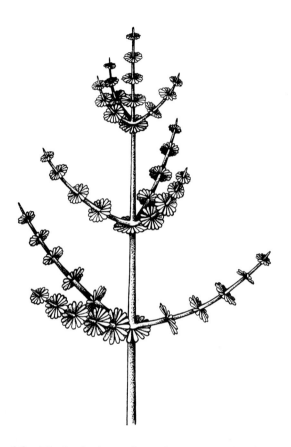

Text-figure 16. The leafy shoot of a *Calamites* tree showing the characteristic whorls of leaves (x 0.5). Drawn by Annette Townsend.

CALAMOSTACHYACEAE

The horsetails reached their maximum physical size and probably their diversity in the Late Carboniferous, with the appearance of the Calamostachyaceae (Text-fig. 15). This group of plants seems to have favoured the mainly wetter parts of the tropical belt, where they flourished in places such as around the fringes of lakes. Their fossils are extremely abundant in the Upper Carboniferous of Europe, North America and China. In the latter area, they continue through into the Permian, but the currently available data is not enough to tell us if they persisted through to the end-Permian extinction, where the giant club mosses finally disappeared.

Text-figure 17. The fertile shoot of a *Calamites* tree showing the characteristic cones (x 0.5). Drawn by Annette Townsend.

Unlike most other horsetails (except *Archaeocalamites*), they were able to develop secondary wood in the stems, which enabled them to grow to heights of at least 10 m. The main trunk arose from a creeping rhizome, and had typical horsetail like nodes that gave rise to branches or, in the higher parts of the plant, leaves. These branches would themselves have borne several orders of branches, producing a tree-sized plant that has been described as looking like a 'giant bottle brush' (Text-fig. 15).

As with most horsetails, the stems had a central ribbed cavity in the pith (Pl. 40), which after death would sometimes become infilled with sediment. The resulting pith cast is usually assigned to the form-genus *Calamites*. The pattern of the ribs, which become offset at each node, indicates that these plants produced the vascular strands to each leaf in a similar way to the living *Equisetum*, and not as in *Archaeocalamites* where there is no offset of the ribs. Another characteristic of *Calamites* pith casts is that they narrow markedly near where they are attached to the lower order branches. Compressions of the outside of the stems are known, albeit less commonly, and these show the scars where branches were attached (Pl. 39). This allows us to reconstruct the general appearance of the whole tree.

Leaves were borne in whorls on the ultimate branches (Text-fig. 16; Pls 41–43). One of the commoner leaf-types belongs to the form-genus *Annularia*, in which there are up to 20 elongate, blade-like leaves in a whorl. Each leaf has a single vein running along its length and has stomata on both surfaces. Some of the species have also been shown to be densely covered with hairs. The leaves were probably borne obliquely to the stem, so as to optimize their angle relative to the prevailing light, thus maximizing their photosynthetic efficiency. This results in the leaf whorls normally found preserved in about the same plane as the stems, in contrast to *Asterophyllites* in which the leaves are often preserved extending into the sediment. *Annularia* leaves are not fused at the base as is seen in the leaf sheafs of modern *Equisetum*. However, close inspection shows that there is a very narrow collar of fused leaf at each whorl, which can be interpreted as an incipient leaf sheaf.

The cones were borne in clusters (Text-fig. 17; Pls 45–46) and have been assigned to several genera, including *Calamostachys* (after which the family and order is named – Pl. 46) *Palaeostachya* (Pl. 44) and *Macrostachya* (Pl. 47). All of the cones have alternating whorls of sterile bracts and sporangiophores with usually two or four sporangia. The different genera are mainly distinguished by the position of the sporangiophores relative to the bracts. For instance, in *Calamostachys* the sporangiophores are attached directly to the cone axis, midway between the bracts, whereas in *Palaeostachya* the sporangiophores occur in the axil of the bract (i.e. in the angle between the bract and the stem). Most calamostachyacean cones are homosporous, but some were heterosporous. One type of cone (*Calamocarpon*) had 'female' sporangia containing just one megaspore and was thus developing towards a seed-like reproductive strategy similar to that seen in some Palaeozoic club mosses (see Chapter 3). The microspores of all calamostachyaceans had elaters as in modern *Equisetum*, but unlike the Archaeocalamitaceae.

GONDWANA HORSETAILS

Horsetails were also prevalent in temperate latitudes during the Late Palaeozoic. The example we have chosen to illustrate this is *Raniganjia* from the Lower Permian of Gondwana (Pl. 48). This superficially resembles the whorls of leaves borne on the Carboniferous calamites from the equatorial belt, but the leaves in each whorl are fused at the base, forming a saucer-like sheaf around the stem. We do not know what reproductive structures were borne on this plant, but the presence of these sheaves around the stem suggest that it might be intermediate between the primitive Calamostachyaceae and the modern Equisetaceae

Another Gondwanan horestail is *Phyllotheca*, which was a modest-sized plant, more on the scale of modern *Equisetum* than the giant *Calamites* of the contemporary tropical floras. Also like modern *Equisetum*, the leaves are fused at their base to form a clear sheaf around the stem. However, the cones have more in common with the Calamostachyaceae, consisting of alternating whorls of bracts and sporangiophores, the latter albeit being more complex (they branch twice to produce four peltate heads, each with four sporangia).

Recent work in South Africa suggests that *Phyllotheca* grew as dense, single-species stands around the margins of shallow lakes. They have been interpreted as analogous to modern-day pastures in today's temperate latitudes, and were probably a major food-source for some of the large vertebrate animals that were starting to appear at this time, such as the mamamal-like reptiles, the dicynodonts.

MODERN HORSETAILS

The modern Equisetaceae differ from the Calamostachyaceae in having bractless cones, stems with no secondary growth and leaves that are fused at the base to form prominent leaf-sheafs. Fossils that correspond to the modern type of horsetail are known from the Tertiary and are confidently assigned to the living genus *Equisetum*. However, there are also fossils in the Mesozoic and topmost Palaeozoic that resemble *Equisetum* but which are not well enough known or show subtle differences from the living forms, and these are called *Equisetites*.

Examples of such a stem are shown in Plate 53. This has well developed leaf-sheafs as in *Equisetum*, but the stems are uncharacteristically large suggesting that there may have been some secondary growth. They are also sometimes associated with cones in which there are occasional whorls of sterile bracts separating zones where there are only whorls of sporangiophores. These factors have led some authors to question whether in fact the Calamostachyaceae should be regarded as a separate family; it should also not be forgotten that some of the Palaeozoic leaf whorls seem to show a narrow collar of fusion at the base of the leaves, although this is nowhere near as prominent as the leaf sheaf of *Equisetum* and *Equisetites*. This may simply reflect the fact that groups of taxa that we call families are man-made constructs and that the fossil record is bound to reveal gradations between what at first sight might seem like sharply delineated groups.

The origin of the modern Equisetaceae has been the subject of considerable debate, with different palaeobotanists arguing for the Archaeocalamitaceae and the Calamostachyaceae as its immediate ancestors. The absence of bracts in the archaeocalamite cones has in particular been used to argue that they were

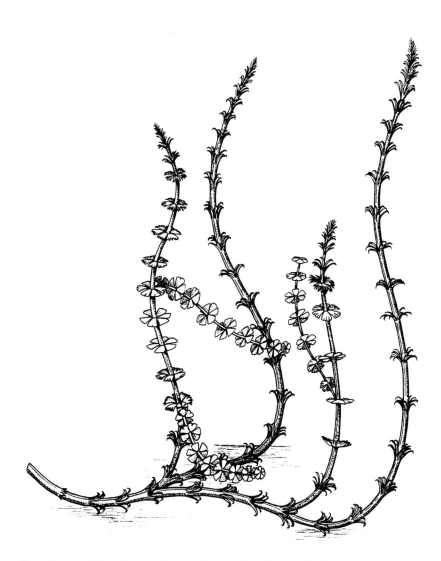

Text-figure 18. The small, scrambling plant *Sphenophyllum* that grew over areas of open ground within the Late Carboniferous tropical swamp forests (x 0.5). Drawn by Annette Townsend, based on the work of L. Batenburg.

ancestors of the modern family. However, emphasis has recently been given to the presence or absence of elaters on the spores, which are thought unlikely to have evolved independently more than once. As they are present in the Calamostachyaceae and Equisetaceae, but absent in the Archaeocalamitaceae, it seems most likely that the first of these families represents the direct ancestors of modern horsetails.

SPHENOPHYLLALES

The last group of plants that we will deal with in this chapter first appeared in the Early Carboniferous or possibly very late Devonian, flourished in the Late Carboniferous (especially in the tropical vegetation) and declined in the Permian. They were herbaceous scrambling plants that probably occupied open, disturbed ground (Text-fig. 18; Pl. 51), and thus filled a similar niche to the goosegrasses and bedstraws (*Galium*) in today's temperate floras. The most commonly found form, at least in the Late Carboniferous tropical floras, is *Sphenophyllum* as illustrated in Plates 50 and 52; one shows a typical European species, while the other represents a Chinese example from the Lower Permian.

The Sphenophyllales have many horsetail-like features: the stems were ribbed, the leaves were borne in whorls, and the cones consist of whorls of sporangiophores and bracts. However, the anatomy of the stems is quite different, having a solid stele usually with a triangular cross section, and no pith cast. The leaves are more complex than those of any other horsetail, having a wedge-shape with numerous dichotomous veins. Also, unlike the Calamostachyaceae, the bracts in the cones are an integral part of the sporangiophores, representing an expansion of the sterile basal part, rather than being independent structures. These distinctive features have caused some authors to question whether the Equisetales and Sphenophyllales share a common ancestor among the horsetails, or whether they evolved independently from the cladoxylaleans. There have even been suggestions that the Sphenophyllales have more in common with the club mosses than the horsetails. Whatever the eventual result of this debate, there can be no doubt that the sphenophyllaleans were important components of the Late Palaeozoic tropical vegetation.

Chapter Five

FERNS

Ferns are the most conspicuous, diverse and widely distributed group of spore-producing plants. They are, in fact, the largest and most diverse group of vascular plant after the flowering plants. Ferns are characterized by having large leaves (fronds) that evolved from planated and laminated branching systems, rather than as stem outgrowths like the microphylls of the lycophytes and horsetails. The great diversity in growth form and life histories has enabled the group to be a successful component of most types of terrestrial plant communities and occasionally to become the dominant component.

The majority of the ferns have been homosporous although a few groups did evolve heterospory. The sporangia are usually aggregated into clusters, called sori, on the underside of the ultimate segments of divided fronds, the pinnules. Many ferns have also developed specialized growth patterns, which enable them to spread vegetatively over considerable areas, while others can reproduce asexually by means of bulbils.

Ferns also constitute the group of vascular plants with the most complex evolutionary history. The fossil record reveals several extensive systematic changes of ferns through geological time, contrasting with the more gradual changes shown by the lycophytes and horsetails. As we will see the earliest Devonian ferns were replaced in the Early Carboniferous by different families which themselves went into decline during the Permian. Some families of extant ferns had their beginnings in the Late Palaeozoic, although the most diverse families evolved and diversified in the Cretaceous and Tertiary.

THE FIRST FERNS

The first fern-like plants appeared towards the end of the Early Devonian, originating as a group of plants with new structural and reproductive characters. At first, all Devonian plants with large dissected leaves were thought to be ancestral ferns until it was discovered that many were in fact progymnosperms (see Chapter 2).

Many of these early ferns are only known from small fragments and there are the usual problems of interpreting the morphology of the whole plant. Nevertheless, we do know that by the middle Devonian two groups of ferns had evolved, the Cladoxylales and the Rhacophytales, and that they retained their identity through to the Early Carboniferous. Both were prominent members of these early plant communities and fossilised remains of the Rhacophytales form the bulk of the earliest coals. The Cladoxylales had a characteristic branching, while a semi-arborescent habit has been proposed for *Pseudosporochnus* and *Calamophyton* (Pl. 54). They have a dissected or very deeply lobed xylem, which often give off strands into the stem appendages, sometimes originating from different lobes of the stele. The small dichotomising ultimate appendages are interpreted as leaves.

Two other groups, the Zygopteridaceae (Pls 55–57) and the Stauropteridaceae (Pl. 58) appeared in the late Devonian although neither formed dominant components of the vegetation. Most of the Zygopteridaceae have small solid vascular strands, although some have larger ones with much internal parenchyma. Within this group there is a tendency towards aggregation of the sporangia into sori on the underside of reduced pinnules. The stauropterids were the first ferns to be heterosporous.

All four groups of early ferns had become extinct by the end of the Palaeozoic, but by then they had been replaced by many other types of ferns that showed a much greater diversity of form.

MODERN FERNS

Modern ferns are usually split into two main groups on sporangial characters. In one group, consisting of the Marattiales and the Ophioglossales, the sporangia develop from several cells. Mature sporangia are relatively large, thick walled and contain large numbers (sometimes over 1000) of spores. They are called eusporangia. The other much larger group of filicalean ferns has smaller, thin walled sporangia that develop from only one cell. These leptosporangia contain relatively few spores (often 64).

There are a number of families of extant taxa that have long fossil histories and there are a few families consisting entirely of rather unusual extinct taxa. Some of the extant families had their origins either in the Late Palaeozoic or Early Mesozoic and their appearance was usually followed by rapid diversification. Certainly by the end of the Palaeozoic there were ferns with the form of rhizomatous herbs, epiphytic herbs, shrubs, vines and tree ferns.

MARATTIALES

The Marattiales was the first modern group of ferns to become a major part of the vegetation. At first these tree ferns were common but not abundant members of the Coal Measures flora, being mainly restricted to drier habitats such as the raised river levees. Towards the end of the Carboniferous, however, they had replaced the arborescent lycophytes as the dominant trees in the back swamps of the forests, at least in Europe and North America. Many of the coals found in these higher strata are the remains of peat produced by these ferns. The marattialeans reached tree size (up to 10 m high) by what has been described as gigantism (Text-fig. 19). They grew vertically with only a little increase in size of their apices and no secondary growth. They were supported by a thick mantle of adventitious roots that produced trunks up to about 1 m in diameter. The stems, when found as permineralizations are referred to the genus *Psaronius* (Pl. 63). The fronds expanded from croziers at the apex of the stem (Pl. 61) into much divided leaves, which may have been up to 3 m or more in length. When the frond eventually dies and is shed from the plant, it leaves a characteristic scar on the stem. Such stems with leaf scars, when preserved as adpressions or casts, are assigned to various form-genera depending on the shape and distribution of the scars; one example is *Caulopteris*, as shown on Plate 62.

The Coal Measures marattialeans are known as both permineralizations and adpressions, which has at times led to a dual system of names. The large fronds disintegrated either prior to abscission or fossilization so they are only preserved in a fragmentary way. The fronds mostly belong to *Cyathocarpus*, *Polymorphopteris* and *Lobatopteris*, which are distinguished generically on a combination of morphological characters. Most can be recognized by having more or less unlobed pinnules. Many of the frond fragments are fertile showing that they produced clusters (sori) of large sporangia on their underside. These are called *Scolecopteris*, *Acitheca* or *Asterotheca* (Pl. 60) Therefore, by careful comparative observation the fronds can be reconstructed with a reasonable degree of accuracy.

Cyathocarpus arborescens is one of the better known species of Late Carboniferous marattialean fronds (e.g. Pl. 59). The fronds of this tree-fern were large, up to 2–3 m long, but with small, tooth-shaped pinnules that had a simple venation. The sori consisted of usually four relatively large, elongate sporangia attached to the underside of the pinnules in two rows on either side of the midvein. As with living marattialeans, the sporangia dehisced longitudinally along a zone of thin-walled cells, to release its numerous spores (unlike most other leptosporangiate ferns, there was not a thickened zone of cells, the annulus, along which the sporangium split). The spores were small, about 20 µm in diameter, trilete, and ornamented with minute lumps (punctae). The remains of this marattialean fern occur abundantly in the uppermost Westphalian and Stephanian of Europe and North America, and it was probably one of the main peat-producers in the late Stephanian.

Ferns

Text-figure 19. One of the tree-ferns which grew within the Late Carboniferous tropical swamp forests and which has a trunk known as *Psaronius*. Redrawn by Howell Reynolds from the work of J. Morgan.

Lobatopteris is another genus of Late Carboniferous ferns, which differs from *Cyathocarpus* in having larger, more elongate and often deeply lobed (pinnatifid) pinnules (Pl. 60). The fructifications were very similar to those of *Cyathocarpus* and there is no doubt that *Lobatopteris* was also a marattialean. The distinctive pinnatifid pinnules occur abundantly as adpressions in the Late Carboniferous and different species occur at different stratigraphical levels, making them a useful type of fossil for dating these strata. In contrast to *Cyathocarpus*, however, *Lobatopteris* never became a dominant part of the back swamp vegetation, always keeping to the drier habitats.

The marattialeans continued to diversify until the Jurassic when the group reached its climax in numbers and diversity. Some fossils from the Middle Jurassic Yorkshire flora are so like extant taxa that they have been referred to the genera of living ferns *Marattia* and *Angiopteris*. The family then steadily declined in number and prominence and there are only a small number of tropical species alive today. Although some species still have very large fronds, none have trunks.

OPHIOGLOSSALES

This is a small order of homosporous, eusporangiate ferns made up of three genera of extant species – *Ophioglossum*, *Botrychium* and *Helminthostachys*. Although the order has been traditionally included with the Marattiales in a loose grouping of eusporangiate ferns and said to be of great antiquity, its origins are, as yet, completely unknown. For many years, its entire fossil record was based upon rather dubious records of dispersed spores. However, fragments of vegetative and fertile fronds have since been found in the Palaeocene of central Alberta (Canada) and these establish an indisputable, although still fairly short, fossil record for the Ophioglossales. These fossils have similarities in structure of both pinnately dissected vegetative fronds and fertile spikes (sorophores) with the extant *Botrychium virginianum*, which interestingly still grows in Alberta today. The fact that these very modern-looking fossils are found at the beginning of the Tertiary does suggest that the order had earlier origins, so we await further discoveries that may throw light on their evolutionary history.

LATE PALAEOZOIC HERBACEOUS FERNS

There were other much smaller herbaceous ferns living in the Coal Measures swamps. They usually have more incised pinnules than those of the marattialeans and a variety of fertile structures which are used to separate such genera as *Renaultia* and *Zeilleria* (Pl. 65). Their affinities are still a matter of some conjecture, but they probably belong to various extinct families such as the Botryopteridaceae, Zygopteridaceae, Crossothecaceae and Urnatopteridaceae.

PALAEOZOIC FILICALEAN FERNS

These leptosporangiate ferns originated in the Carboniferous and recent evidence from fertile filicalean frond segments from eastern North America suggests that they might even have first evolved near the start of the Carboniferous.

One of the best known examples of a Carboniferous leptosporangiate fern is *Pecopteris plumosa*, whose fronds occur abundantly in the Upper Carboniferous Coal Measures of Europe (Pl. 64). It was a tree-fern, superficially very similar to the Carboniferous marattialeans dealt with above, with large fronds bearing small, dentate pinnules. However, the fertile structures were quite different from the marattialeans. The fertile pinnules bore sporangia, singly or in loose clusters, near the margin and associated with veins. They have a clear apical annulus, which at one stage was regarded as evidence that these ferns belonged to the extant family, the Schizeaceae. However, there are a number of other features of the plant as a whole, such as the presence of small, vascularized scales on the stem, which are not found in the Schizeaceae. Furthermore, *Ankyropteris brongniartii*, an anatomically preserved fern that is closely related to (if not conspecific with) *P. plumosa*, has been shown to have axillary branching, a feature that is very unusual in ferns and is more normally associated with seed-plants. The consensus now seems to be that *P. plumosa* is a primitive leptosporangiate fern, belonging to extinct Palaeozoic family the Tedeleaceae.

OSMUNDACEAE

Many groups of ferns appeared after the severe climatic changes at the end of the Palaeozoic. The primitive family, the Osmundaceae, which appeared in the Late Permian of both the Northern and Southern Hemispheres, has the most extensive fossil record of any family of ferns. The family is now known from more than 150 species of fossils: most are adpressions, although about 50 are permineralized stems and 20 are isolated spores. Today there are three genera with 21 living species. One distinctive character of living taxa is that each species has two different and quite distinctive types of frond: vegetative fronds in which the pinnules have a broad lamina for photosynthesis, and fertile fronds in which the sporangia-bearing segments have very little if any lamina.

The structural diversity of the fern axes and the widespread distribution of fossil and living genera and species indicates that the group originated in the Early Permian or even earlier. Evolution was then rapid during the Late Palaeozoic and Early Mesozoic, and by the Jurassic there were several marked similarities between the plants living then and those living today. Even though their rhizomes and petioles can be closely compared with that of living members of the Osmundaceae (Pl. 66), it is often much more difficult to be so certain about foliage identification unless it is fertile. Nevertheless, some leaves have been assigned to living

Text-figure 20. *Weichselia* was a Mesozoic fern that spread vegetatively to cover large areas of land to form a savannah. The fronds were about 1 m long. Drawn by Annette Townsend, based on the work by K. Alvin.

genera where they are nearly identical to the foliage of living taxa. Evidence of such fertile foliage shows us that the family had developed its habit of dimorphic fronds prior to, or during, the Jurassic (Pl. 67). *Todites* and *Cladophlebis* are both common Mesozoic members of the Osmundaceae.

SCHIZAEACEAE

The family today consists of four tropical or subtropical genera and about 160 species. Most species are creepers or climbers. Their sporangia are borne singly, are large and have a very primitive type of dehiscence with terminal specialized cells (annulus).

The family's origin seems to be in the Triassic. Early members include the genus *Klukia*, which is common in floras of Late Triassic, Jurassic and Cretaceous ages. There are also fossils that can be assigned to extant genera, such *Anemia* from the Lower Cretaceous and *Lygodium* from the Lower Tertiary (Pl. 70).

MATONIACEAE

Members of the family Matoniaceae first appeared in the Late Triassic of the Western United States. These fossils, belonging to the genus *Phlebopteris*, are both abundant and well preserved (Pl. 71). The plants had palmate leaves with an odd number (5–15) of spreading pinnatifid pinnae. The sori are in single rows on either side of the pinnule midrib and contain 14–20 sporangia each with an oblique annulus. Reconstructions suggest that the leaves were produced by horizontally growing rhizomes. All these characters clearly suggest affinity with living members of the Matoniaceae. However, because the *Phlebopteris* fossils have so many matoniaceous characters, it is likely that plants with at least some of the characters existed earlier. By the Jurassic, matoniaceous ferns were widespread around the world and they continued to be so in the Cretaceous with such successful genera as *Weichselia* forming dense swards by rhizomatous growth, rather like modern day bracken (Text-fig. 20; Pl. 69). Even so, climatic change in the Middle Cretaceous in the Northern Hemisphere brought havoc to the family which seems to have died out there by the Late Cretaceous. Indeed there are no known matoniacean fossils of Late Cretaceous or Tertiary age and yet, sometime during this time-span, the genus survived and became restricted to only two living genera in the Malaysian-Borneo area.

DIPTERIDACEAE

The family contains only eight extant species that are restricted to the Indo-Malaysian-Polynesian region and seven Mesozoic species from both hemispheres. The leaf architecture is quite distinctive in developing through a succession unequal dichotomies that produce a flattened, almost umbrella-like appearance. The members of the Dipteridaceae have a very similar pattern of distribution in time and space to those belonging to the Matoniaceae, which suggests a close relationship between the two families. The origin of both appears to have been in the Early Triassic or the Late Palaeozoic and both became world-wide in distribution during the Mesozoic. Today they have a similar distribution.

DICKSONIACEAE

This family contains five genera of Southern Hemisphere tropical and subtropical tree ferns. The most diagnostic character is the marginal to submarginal sorus which is protected by a bivalved structure, the valve nearest the midvein consisting of a cup-shaped membrane (known as an indusium) and the more marginal valve formed by the reflexed margin of the leaf lamina.

The family had probably originated by the Early Triassic because dicksoniaceous ferns are so abundant in some Jurassic and Lower Cretaceous floras. The two best known genera are *Coniopteris* and *Dicksonia*, which appeared in the Jurassic. *Coniopteris* fronds, when fertile, have sori enclosed in cup-shaped indusia on the ends of reduced pinnule segments.

The family became gradually restricted to the Southern Hemisphere during the Late Cretaceous, and only a very few definite fossils representatives of Tertiary age have been found in the Northern Hemisphere.

POLYPODIACEOUS FERNS

The family Polypodiaceae has been constantly changed in content and often divided into many, much more precise, families. The only characters that are constant throughout the Polypodiaceae in its broad sense relate to the sporangia. They are small, usually stalked, have a vertical, incomplete annulus and usually contain 64 spores.

Several Mesozoic fossils have been referred to the broad Polypodiaceae and included in genera of extant ferns such as *Aspidium, Asplenium, Davallia, Polypodium* and *Pteris*. However, the major problem in evaluating and accepting these claims is that nearly all of them are sterile foliage. There are very few specimens that can be referred to the family with any certainty. The earliest example that might possibly be acceptable is *Aspidistes thomasii* from the Yorkshire Jurassic, but even here there is no unequivocal proof.

There are a number of Tertiary fossil ferns that are definitely polypodiaceous, although there are others which are only known from portions of sterile foliage and therefore very debatable in affinity.

Fragmentary specimens from Cretaceous and Tertiary rocks have been assigned to the genus *Onoclea* which is regarded as belonging to the most advanced group of filicalean ferns, the dennstaedteoid-aspleniod line. The best published example of an undisputed fossil *Onoclea* has even been referred to the living species *Onoclea sensibilis* that now grows throughout northern Asia and central and eastern North America. Thousands of specimens were recovered from Palaeocene non-marine flood deposits in Alberta, Canada. They included large segments of whole plants in positions of growth, within a flora dominated by a birch relative (*Paleocarpinus*) and the Dawn Redwood (*Metasequoia*). Today, *O. sensibilis* grows in moist woodlands and along the sides of streams and marshy lakes.

Other undisputed fossil ferns are much less spectacular. *Achrostichum preaureum* has recognizable sporangia and dispersed sporangial masses, and isolated sporangia from the British Tertiary have been referred to *Achrostichum anglicum*. Occasionally anatomically preserved rhizomes can be closely compared with those of extant ferns allowing them to be identified with reasonable certainty. Rhizomes of *Dennstaedtiopsis* have been described from Oregon and these *Histiopteris* from the London Clay of south-east England.

The paucity of fossil ferns does not necessarily mean that polypodiaceous ferns were virtually absent in the Tertiary or even in the Cretaceous. It is most likely that they are simply under-represented in the fossil record. Most polypodiaceous ferns have fronds that die, collapse and wilt on the plants. In contrast, flowering plants and conifers shed their leaves in vast numbers so Tertiary leaf floras, in particular, must give and over-representation of flowering plants and an under-representation of ferns.

TEMPSKYA

Although most Mesozoic fern fossils can be accommodated within the modern fern families, one rather bizarre type is quite different from anything alive today: the tree fern *Tempskya*. It is the only known genus of the family Tempskyaceae, and grew throughout North America, Europe and Japan in the Cretaceous. The plants grew upwards forming a false trunk of several intertwining stems and roots (Text-fig. 21). All the leaves were small and spread over the upper portions of the trunks unlike the large leaves of other tree ferns that grow in an apical crown. We only know of this fern through petrified portions of the false trunk and associated leaf bases and roots. Nothing is known of the reproductive structures that it produced. However, its mode of growth is so different from that of any other known fern, either alive or extinct, that there is general agreement that it must belong to its own family.

Text-figure 21. The tree-fern *Tempskya* had a trunk of interlocking thin, rhizomatous stems embedded in adventitious roots (x 0.03). Thin, delicate leaves emerged over the length of the fire-resistant trunk. Drawn by Annette Townsend, based on the work of H. N. Andrews and E. M. Kern.

HETEROSPOROUS FERNS

There are two orders of extant heterosporous ferns. Both produce their megaspores and microspores in sporangia that are encased in desiccation resistant structures called sporocarps. This common feature indicates that both orders probably had their origins in ancestors that grew in seasonal pools.

The Marsileales contains three genera of rhizomatous, rooting ferns that often grow in dense mats (*Marsilea*, *Regnellidium* and *Pilularia*). The sporocarps are rather like a small pea pod in appearance and are attached at the base of a frond. The fossil record of this order is sparse with only a few debatable megaspores described from the Cretaceous and the Tertiary.

The Salviniales contains two genera, *Salvinia* and *Azolla*. Both are small and free floating on the surface of lakes and very slow moving streams. They have horizontal leafy stems and their sporocarps hang downwards. *Azolla* has the larger fossil record because the complex reproductive structures from its sporocarps are commonly found preserved (e.g. Pl. 72). The megaspores have floats derived from other aborted megaspores that are also impregnated with sporopollenin. The number of floats is an important character in distinguishing fossil species of *Azolla*. Clumps of microspores, called massulae are often found attached to the megaspores as they are in extant species. The oldest known megaspores referable to *Azolla* are latest Cretaceous and these had up to 24 floats. The numbers of floats appears to decline in younger strata and *Azolla* species are placed in one of three groups based on float numbers: one to three floats, four to nine floats and over nine floats. As yet there is no clear idea of the evolutionary significance of float number reduction.

The oldest vegetative remains of *Azolla* plants are Palaeocene fossils from Alberta, Saskatchewan and South Dakota in North America. The plants are up to 22.5 mm long with alternate, imbricated leaves along the branched stems. Some specimens are deduced to have been fertile on the basis of closely associated megaspores, some with attached massulae, and separate massulae. The sporocarps appear to have decayed prior to preservation. These specimens indicate that most of the vegetative characters of modern *Azolla* had evolved by the Middle Palaeocene.

Chapter Six

EARLY SEED PLANTS

After plants became established on land, the next major development in their evolution was the appearance of the ovule. As long as plants had an independent gametophyte, in which fertilization took place in an unprotected environment, vegetation would be more or less tied to the damper terrestrial habitats. The ovule brought several advantages, but the most significant was that it provided a protected environment for fertilization to take place and seeds to be formed. This released plants from their dependency on external moisture.

There are two major types of seed plants, the gymnosperms and angiosperms, distinguished mainly on whether or not the seeds are borne naked or enclosed in an ovary. The gymnosperms, which have naked seeds, are the subject of this and the following chapter. Gymnosperms arose in the Late Devonian and diversified significantly during the rest of the Palaeozoic. Many of the Palaeozoic gymnosperm groups became extinct at the P/T (Permian – Triassic) boundary, to be replaced by other groups such as conifers, ginkgos and cycads. During the Mesozoic, gymnosperms dominated most land vegetation, with only the ferns proving a match for them among the pteridophytic plants. Towards the end of the Mesozoic, however, they became progressively replaced by angiosperms. The K/T (Cretaceous – Tertiary) event, which saw the extinction of the dinosaurs, also resulted in a number of gymnosperm groups becoming extinct, but others (most notably the conifers) survived and continued to be important components of land vegetation right through to the present day.

There is such an abundance and diversity of gymnosperms in the fossil record, that it is impossible to deal with them in a single chapter. We have therefore divided them into 'Early gymnosperms' (this chapter) and 'Modern gymnosperms' (Chapter 7). The division is entirely artificial and no systematic or botanical meaning should be read into it. The 'early' forms are mainly those found in the Palaeozoic and which became extinct at the P/T boundary. Some of the 'modern' gymnosperms first arose in the Palaeozoic (e.g. cycads and conifers) but came into their own during the Mesozoic.

WHAT ARE OVULES AND SEEDS?

An ovule is the female reproductive organ of a seed-plant, before it becomes fertilized. It represents the ultimate evolutionary result of heterospory, in which the female sporangium contains just a single megaspore. The wall of the sporangium in an ovule is normally referred to as the nucellus and the megaspore wall as the embryo sac.

Around the nucellus is a second layer of protective tissue. In the earliest ovules (sometimes referred to as pre-ovules) this protective layer consists of a sheaf of sterile axes, but in true ovules, this sheaf became fused to form a continuous layer of tissue, known as an integument. Many ovules also have yet another protective layer called the cupule. Cupules probably evolved in an essentially similar way to the integument but, unlike the integument, there may be several ovules within a single cupule.

In the distal part of true ovules, there is an opening (micropyle) in the integument to facilitate the capture of pollen. Most living and probably most fossil gymnosperms produced a liquid which bulged out of the micropyle, to capture pollen carried in the wind. The bulging pollen drop is reabsorbed, taking in the pollen which passes through the micropyle and enters a cavity known as a pollen chamber. In living cycads and probably the early gymnosperms, the male gametophyte liberates motile gametes (sperm). These swim through the remaining pollen drop to the archegonia and enter through the neck exposed at the top of the embryo sac. Here the pollen germinates to produce the male gametophyte. In the more advanced gymnosperms, the conifers, this gametophyte is itself only a minute structure consisting of just a few cells. Such microgametophytes remain within the wall of the pollen grain, producing a pollen tube along which the male nuclei pass to the archegonia. In some of the early gymnosperms, the pollen retains some features of the ancestral microspores, such as the trilete mark on the surface. However, it is assumed that, like modern gymnosperm pollen, they did not produce free-living gametophytes and is therefore called pre-pollen.

Pre-ovules, because they do not have a proper integument, obviously do not have a micropyle either. Instead, the distal end of the nucellus is modified into a trumpet-shaped structure (a lagenostome) that fulfils a similar function to the micropyle, to enable pollen-capture.

On fertilization, the ovule becomes a seed and is released from the parent plant, to grow into a new individual. In many of these plants, the seed has features that help in its dispersal, such as wings in wind-dispersed plants, or develops a fleshy outer coat for those plants using animals as the dispersal vector.

In many modern seed-plants, there is a time after fertilization when the embryo lies dormant in the seed, often awaiting particular environmental change to trigger germination (e.g. a period of winter frosts). However, one of the curiosities of the fossil record is that very few seeds of the early gymnosperms are found with embryos. This cannot be a problem of poor preservation, because many examples

Text-figure 22. *Elkinsia*, one of the earliest known seed plants, from the Late Devonian of North America (x 0.15). The foliage consisted of highly segmented fronds, while the ovules were borne in clusters at the top of the plant. Drawn by Annette Townsend, based on the work of G. W. Rothwell, S. E. Scheckler and W. H. Gillespie.

are known where the much more delicate gametophyte can be recognized. It has been suggested that very few primitive gymnosperm ovules were fertilized, but it seems more likely that they simply did not have a significant dormancy period between fertilization and germination, because it was an adaptation that did not appear until much later in the evolutionary history of gymnosperms.

ELKINSIACEANS

The most primitive order of seed plants is known as the Lagenostomales, in which there are two well documented families. Fossilized remains of the oldest of these families, the Elkinsiaceae, have been described from the Upper Devonian (Fammenian) of Europe and more especially North America (West Virginia). Most are isolated fragments of leaf or seed, but one plant of this age has been almost completely reconstructed: *Elkinsia polymorpha* from the Upper Hampshire Formation of West Virginia (Text-fig. 22). It was an herbaceous plant, probably no more than 1 m high, with a slender stem bearing helically-arranged, large fronds. The stems had a stele with a trilobed cross-section and there was some secondary wood in the thicker stems. As with many Palaeozoic gymnosperms, the vegetative fronds were highly dissected, superficially resembling fern-fronds. This apparent resemblance caused early authors to refer to such plants as seed-ferns or pteridosperms. These are unfortunate names, as the plants have little to do with true ferns, but the names have stuck and are still regularly found in the palaeobotanical literature to describe the early gymnosperms with large-fronds.

The lower part of the *Elkinsia* fronds had a major dichotomous fork of the main rachis, a feature seen in most Palaeozoic pteridosperms. Each branch produced by the fork was further divided by two orders of lateral, pinnate branching, the last order bearing narrowly lobed leaflets or pinnules.

Unlike many later pteridosperms, the reproductive structures were borne on fertile fronds that were quite different from the vegetative fronds. The fertile fronds were not planated, but three-dimensional branching systems, in which either cupulate pre-ovules or pre-pollen bearing structures were borne at the ends of the branches. Although the evidence is not conclusive, it is likely that *Elkinsia* had separate male and female fronds. The general form of these fertile fronds has been interpreted as an adaptation to wind fertilization.

In the female fronds, four pre-ovules were borne within each cupule, the latter consisting of sixteen free, elongate segments. The pre-ovules have a pre-integument consisting of four to five elongate lobes, that become fused to each other and the nucellus in the lower half of the pre-ovule, and a lagenostome typical of the earliest known gymnosperms.

Fragments of similar pteridosperms have been reported from Late Devonian strata in Britain (Devon), Ireland and Belgium (e.g. Pl. 74), suggesting that the gymnosperms arose in what was then the tropical latitudes. However, their remains

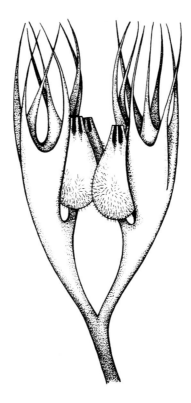

Text-figure 23. The female reproductive structure of the early gymnosperm *Archaeosperma*, showing four ovules borne in a protective cupule (x 5). Drawn by Annette Townsend, based on the work of J. Pettitt and C. B. Beck.

do not become abundant in the fossil record until the Lower Carboniferous, which is when we see the first real diversification of this group of plants.

LAGENOSTOMACEANS

During the Early Carboniferous, the Elkinsiaceae became progressively replaced by another, more advanced family of the Lagenostomales, the Lagenostomaceae. This order became particularly abundant in the tropical vegetation of the late Early and early Late Carboniferous, eventually becoming extinct in the early Stephanian. It had stems with a simple central stele, although some of its component species, especially in the Early Carboniferous, had substantial amounts of secondary wood and were probably quite large plants. When first found, such wood was thought to be coniferous and assigned to the form-genus *Pitus*. However, true

Text-figure 24. Although most of the Late Carboniferous pteridosperms are believed to have been trees, one group of lagenostomaleans (*Mariopteris*) was probably a liana that grew up the trunks of other trees. Drawn by Annette Townsend.

conifers do not appear in the fossil record until much later (see Chapter 7) and it has since been shown that the Early Carboniferous wood belonged to plants that produced pteridosperm-type foliage.

The lagenostomalean foliage was similar in many ways to that of the Elkinsiaceae (e.g. Pls 77–79). There was a major fork of the rachis in the lower part of the frond, producing two branches that are more or less mirror-images of each other. The ultimate segments of the fronds were pinnules with deep, finger-like lobes.

There are, however, many significant differences between the two orders. Most obvious is that, although the Lagenostomaceae bore clusters of cupulate ovules (Text-fig. 23) and pre-pollen bearing synangia, they were not attached to separate fertile fronds. They were instead borne on a short truss of axes which itself was attached to the angle of the main fork of the rachis of otherwise normal-looking fronds, resulting in a trifurcation (Pl. 79). The female reproductive units were also quite different, being true ovules with an integument that entirely enveloped the nucellus and megaspore, except for the micropyle pore at its distal end (Pls 75–76).

There is still a lagenostome at the distal end of the nucellus (hence the name of the order), reflecting the probable ancestry of the family in the Elkinsiaceae, but it is a much reduced structure that was at least partly redundant; the capture of the pre-pollen was now mainly the function of the micropyle.

In the Late Carboniferous, most lagenostomaleans were relatively small plants, none producing the thick stems with substantial secondary wood seen in the Early Carboniferous. One way that plants of this order overcome the problem of lack of height, which would have been a major problem in the dense tropical forests in which they grew, was to become a liana or vine, climbing up the trunk of trees (Text-fig. 24). The best known examples of this belong to the form-genus *Mariopteris* (Pls 80–81). The underlying form of *Mariopteris* fronds does not differ significantly from their Early Carboniferous ancestors, but they were much smaller, mostly less than 0.5 m long. However, they can be readily recognized by their ultimate segments, which have much more rounded lobes. The fronds were helically attached to rope-like stems with an anatomy that seems well adapted to a liana habit. This was confirmed by the discovery of a fossil stem with fronds attached and still wound around the stump of a giant club-moss trunk in the Westphalian of northern England.

MEDULLOSALES

In the Late Carboniferous tropical forests, the large pteridosperms mostly belonged to the Medullosales (sometimes alternatively referred to as the Trigonocarpales). They were mainly small trees with fronds in the order of a metre in length (Text-fig. 25), although one species (*Alethopteris lonchitica*) had substantially larger fronds, at least 7 m long that were presumably borne by a much larger tree.

One of the distinctive features of this group of plants was the stele. Rather than there being a simple, solid stele running along the stem, as in the Lagenostomales, the medullosalean stems appear to have several separate steles, usually with some secondary wood, separated by softer tissue. This was originally interpreted as having been the result of the fusion of several separate stems, with the steles still remaining independent. This type of vascular system, technically known as a polystele, occurs today in tree-ferns. However, a careful analysis of the medullosan stem anatomy has shown that it is in fact not polystellic, but consists of just one stele which has become dissected into a number of strands, which divide and fuse again along the length of the stem. This structure provided the stem with considerable mechanical strength, but without the heavy physiological cost of producing large quantities of 'expensive' secondary wood that would be needed if the stele was a solid structure. This may explain why in the Late Carboniferous the medullosans were able to become such large plants compared to the Lagenostomales.

The success of the medullosans in the Late Carboniferous tropical forests is

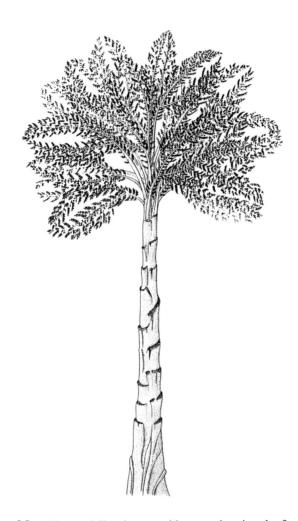

Text-figure 25. The medullosalean pteridosperm bearing the fronds known as *Neuropteris* was a common tree growing on the levees of the Late Carboniferous tropical swamp forests. The size of the fronds varied considerably according to the type of plant; this example had fronds 1–2 m in length. Drawn by Pauline Dean, based on the work of C. J. Cleal and C. H. Shute.

reflected in the range of different types of frond found preserved in the rocks of this age. The largest fronds (reaching lengths of at least 7 m) have elongate, tongue-shaped pinnules and are assigned to the genus *Alethopteris* (Pl. 84). Somewhat smaller fronds, mostly about 1 m long, tend to have pinnules with a constricted base that were traditionally assigned to *Neuropteris* (Pl. 87). However, recent work on the branching patterns of the rachises within the frond, and of their epidermal structures, have allowed the recognition of several new genera, such as *Laveineopteris* and *Macroneuropteris*. In the very late Carboniferous (Stephanian) we see even smaller fronds, only 0.5 m or so long, belonging to *Odontopteris*, and which may have been scrambling plants or lianas. All of the fronds have a basal dichotomy of the main rachis, producing two branches that are three- or four-times divided into fern-like pinnae. Some of the genera also have short, once-divided pinnae attached to the main rachis between the larger, three- or four-times divided pinnae; these are known as intercalated pinnae. Some of the genera, such as *Laveineopteris*, also had large, round leaflet-like appendages attached to the base of the frond, which probably functioned as a means of disposing of excess water in the body of the plant.

Most of the medullosalean pinnules have simple or dichotomous veins, but some also develop a meshed or anastomosed venation. This is, however, a simple form of anastomosed venation, in which the veins have become sinuous to the extent that adjacent ones touch and eventually fuse. For one genus (*Neuropteris*) it is possible to trace, through the sequence of lower and middle Westphalian rocks, a gradual increase in the sinuosity of the veins, culminating eventually in an anastomosed equivalent known as *Reticulopteris*. This simple style of anastomosed veining is quite different from that seen in today's angiosperms, which comprises several different orders of veining.

Medullosalean foliage often has a thick cuticle showing evidence of the underlying epidermal cells (Pl. 88). These cuticles have been the subject of several studies, and the distribution and structure of the stomata have proved important features for helping understand the systematics of these fronds. In some of the fronds, such as *Neuropteris*, one of the distinctive features is the presence of numerous pores near the margin. These are known as hydathodes and were where the plant exuded water from the leaf. Hydathodes are today found in many plants growing in tropical rain forests and are an adaptation to growing in humid conditions.

Seeds and ovules occur reasonably commonly in the fossil record, but almost invariably detached from the parent plant (Pl. 85). We know very little about how they were attached to the plant. There have been recorded examples of individual ovules being found apparently attached to a frond, replacing one of the vegetative pinnules, but it is difficult to be certain that these are not just examples of detached ovules lying on top of frond fragments. The medullosans had some of the largest seeds produced by any gymnosperm, the largest recorded example being 11 cm long. When anatomically preserved (Pl. 86), they show themselves to be almost identical in structure to those of living cycads, with which they are believed to be closely related. They are among the earliest of the modern seeds, in which there is no lagenostome at the distal end of the nucellus, pollen-capture being fully the function of a micropyle. The seeds typically have three ribs running along their length, making them easily recognized even in compressions. Casts of the inside of the nucellus, which are often found in certain sandstone deposits, retain this characteristic tripartite division (Pl. 85).

The pre-pollen found in the pollen-chambers of medullosan seeds is also very cycad-like. They had a simple monolete mark (a single line), rather than the triradiate 'Y' mark of the lagenostomaleans, and they seem to have produced motile male gametes resembling those of the cycads. The pre-pollen was produced by compound synangia, probably attached directly to the fronds, although this again is not totally certain. The synangia consisted of clusters of elongate sporangia fused together along their length. They varied considerably in complexity from clusters of just four to over one thousand sporangia.

The medullosans were very much plants of the Late Carboniferous tropical forests. The earliest known examples are from the topmost Lower Carboniferous, but they do not become abundant until the Upper Carboniferous (especially the Westphalian) and eventually disappear in the Lower Permian. They are mostly known from Europe and North America, but there are some species also known from central Asia and China.

PARISPERMACEANS

Paripteris and *Linopteris* are fronds that bear a striking resemblance to many medullosans, and occur widely in Europe and North America in the Upper Carboniferous (Pls 89–91). However, detailed studies have demonstrated that the fronds had a fundamentally different architecture, such as the absence of the basal main dichotomy. The male fructifications (*Potoniea* – Pl. 92) have also proved to be significantly different, consisting of numerous clusters of four elongate sporangia set in a mass of sterile tissue, and which produced pre-pollen with a triradiate rather than monolete mark. However, the seeds show greater similarity in their basic anatomy, although they have a six- rather than threefold symmetry.

The relationship of these plants to the medullosans has never been properly resolved. However, recent evidence of their distribution now suggests that they were only distantly related. As stated above, the true medullosans are characteristically European and North American species. However, the parispermaceans seem to have first appeared in the Far East (China) in the Late Visean, and only later migrated west to Europe, where they do not appear until the Namurian.

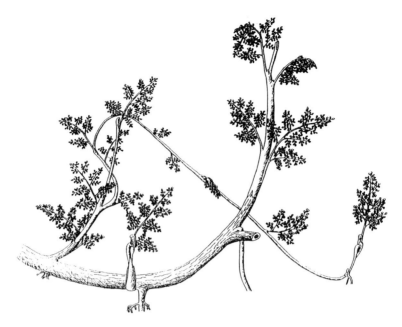

Text-figure 26. The scrambling pteridosperm *Callistophyton*, which grew in the tropical swamp forests towards the end of the Carboniferous (x 0.2). Redrawn by Deborah Spillards from the work of G. W. Rothwell.

CALLISTOPHYTES

This is another group of pteridospermous plants that grew in the Late Carboniferous tropical forests. Although apparently not as common as either the Medullosales or Lagenostomales, they have received considerable attention from palaeobotanists, partly because they appear to share characteristics of several other groups of gymnosperms.

The best studied (the *Callistophyton*-plant) was a small shrub with a rather scrambling habit (Text-fig. 26). The main stems were up to 3 cm wide, with an anatomy that was rather similar to some lagenostomaleans and cordaites, with a central pith around which were several strands of primary xylem. This was then entirely surrounded by secondary wood. The cortex around the secondary wood is very distinctive, in having numerous cavities lined by secretory tissue, thought to have produced resin. Similar cavities occurred in the foliage and reproductive organs, and were one of the means of reconstructing the plant, even though organic connection between many of these organs has never been found.

The best known foliage of this group belongs to the form-genus *Dicksonites*. Like most other pteridosperms, the frond has a near-basal dichotomy of the main rachis. The distinctive pinnules are usually robust and somewhat vaulted.

The reproductive organs were borne directly on the underside of pinnules of normal fronds. The seeds bear some superficial similarities to those of the medullosaleans, having a complex integument that is only fused to the nucellus in its basal part, and no lagenostome. However, the callistophyte seeds (usually referred to as *Callospermarion*) are flattened, with two apical horns. Such flattened (or platyspermic) seeds are normally associated with cordaites and conifers. However, the significance of this symmetry of the seed has recently been questioned as there is now evidence that there is an underlying radial-symmetry in the basal part of the *Callospermarion* seeds, and that its platyspermic symmetry can be derived from the sort of structures that we can see in lagenostomalean seeds.

The pollen-organs are like those of many pteridosperms, consisting of radial clusters of sporangia, although in the callistophytes they are attached directly to the underside of pinnules. This gives such pinnules the superficial appearance of fertile fern foliage, although there is no suggestion that they are closely related. The most distinctive feature of these organs is the pollen, which seems to be true pollen with two bladders presumably to assist in wind dispersal. Such pollen is very similar to that found in cordaites and conifers.

Like the medullosaleans, the callistophytes were very much plants of the Late Carboniferous western tropical forests. They first occur in the lower Westphalian, reach their zenith in the topmost Westphalian and lower Stephanian, and appear to become extinct in the lower Permian. Whether they survived in the Permian tropical forests in China remains to be demonstrated.

PELTASPERMS

As the medullosaleans and callistophytes were becoming extinct in the Early Permian, another group of pteridosperms started to appear. We have little direct evidence as to the size of these plants but, from the size of their leaves and the associated stems, they were probably relatively small shrubs. The ovules were borne on stalked sporangiophores, with several ovules attached to the underside of a mushroom-shaped head. The pollen was produced by cup-shaped clusters of several sporangia that were fused at the base. In some of the earlier forms, the reproductive organs were borne in loose cones, but mostly they were in irregularly branched clusters.

The fronds were small relative to many of the Carboniferous pteridosperms, and could be quite variable in structure. One of the most typical of the Early Permian fronds in Europe is referred to as *Autunia* (formerly *Callipteris*) and most fronds were only just over 0.5 m long (Pl. 93). Unlike most pteridosperms, this does not have a near-basal dichotomy of the main rachis, although in the apical part the fern-like, pinnate branching sometimes breaks down to produce what are called pseudo-dichotomies. There was just one order of lateral branching, and pinnules were attached to both these and the main rachis.

At about the same time, peltasperms with another type of leaf, known as *Supaia*, were growing in China and North America (Pl. 94). These were much smaller fronds than *Autunia*, often less than 20 cm long, and had a major dichotomy of the main rachis. *Supaia* also differs in not having any lateral branching. The form of the leaf is so different from that of *Autunia* that their relationship might be questioned. However, recent work in China has revealed ovuliferous cones associated with *Supaia* leaves, which are very similar to those of *Autunia*.

Although not illustrated here, there are well-documented peltasperms found also in middle and high latitude Permian floras of both Northern and Southern Hemispheres. The peltasperms are also known from Triassic floras and clearly survived the Permian-Triassic extinction event, where so many of the Palaeozoic gymnosperms succumbed. This is all a clear testimony to the adaptability of the peltasperms, especially to the somewhat drier habitats represented by the Permian and Triassic floras. Why, then, did they disappear in the Mesozoic? Perhaps they could not adapt to the heavy grazing by herbaceous dinosaurs that were starting to appear to this time. Alternatively, they may not have disappeared, as such, but simply evolved into one of the better-known Mesozoic gymnosperm groups such as the Umkomasiaceae.

GIGANTOPTERIDS

At the same time that the peltasperms were abundant in parts of China and North America, another group of early gymnosperms was present there. They are known informally as gigantopterids (Pls 95–96). Their leaves are extremely common in the Permian of the Far East, and are regarded as one of the characteristic members of what is known as the Cathaysian Flora. They can be fronds or entire leaves, with the former showing pinnate and/or dichotomous branching. The unifying character is the venation, which consists of several orders of vein-branches, the ultimate orders coalescing to form a reticulate venation. This remarkably advanced-looking venation seems rather like that found in many Recent dicotyledon leaves and at one time the gigantopterids were considered as possible ancestors of the angiosperms, although the view is now discounted.

The botanical affinities of these plants have long been regarded as problematic because of the very limited evidence that there is available concerning their fructifications. They tend to be placed in their own order (Gigantonomiales) but how this relates to other gymnosperms is ambiguous. Even the general habit of the plants that produced these leaves is uncertain. Some Chinese palaeobotanists suggested that they were lianas clambering up the trunks of the Permian tropical forests, but it has been pointed out that the fossil assemblages where the gigantopterids occur do not include the remains of any trees. Recent unpublished work on their cuticles has suggested that some were partly submerged, aquatic plants, while others were adapted to relatively drier habitats. However, it is clear that much remains to be worked-out about the plants that produced these widely-distributed leaves.

GLOSSOPTERIDS

Rocks of Permian age in South America, southern Africa, India and Australia often contain abundant remains of leaves (e.g. Pls 97–98) belonging to the general group of plants known as the glossopterids (some now argue that they are better referred to as the Arberiales, but here we will continue to use the traditional, informal name). The presence of this glossopterid foliage in such widely separated locations and at such different latitudes was a puzzle to the nineteenth century palaeobotanists. In the 1920s, however, the German climatologist Alfred Wegener gave the novel explanation that these areas all lay at much higher latitudes in the Late Palaeozoic than they do today, and had been joined together to form part of a super-continent known as Gondwanaland. The distribution of the glossopterids became one of the most important pieces of evidence supporting the theory of continetal drift and remained so until the discovery of plate tectonics, which provided a mechanism to explain the movement of the continents. Today, this type of evidence from plant fossil distribution remains one of the best means of judging the past positions of continents and for creating palaeogeographical maps.

Glossopterid leaves were entire, not divided like those of the pteridosperms, although one cycad-like leaf has been reported from the Lower Permian of India that has a glossopterid epidermal structure (*Pteronilssonia*). There can be considerable variation in leaf-size, the largest entire-margined leaves reaching more than 30 cm in length. The most commonly found genus, *Glossopteris*, after which the group is informally named, is characterized by a strong midvein running along its length and anastomosed lateral veins. These anastomosed veins are quite different from the venation of the gigantopterids dealt with previously, because they do not consist of several different order of veins, and are instead more similar to what is seen in the medullosaleans. Other, less abundant leaf-types also probably belong to the glossopterids, such as *Gangamopteris* which does not have a midvein, and *Rhabdotaenia* which has non-anastomosed veins.

The leaves were almost certainly produced by substantial-sized trees (Text-fig. 27), although this idea is partly speculative since the leaves are mostly found detached from the stems. The frequency with which detached leaves are found has led some authors to suggest that they were abscised in the autumn, as do most angiosperms living today in temperate climates, but no specialized abscission layer has yet been identified. Wood sometimes associated with the leaves, which looks rather like conifer wood with prominent growth rings, is generally assumed to be the trunk and branches of the tree that bore them.

Several different types of reproductive organ (e.g. Pl. 97) have been found

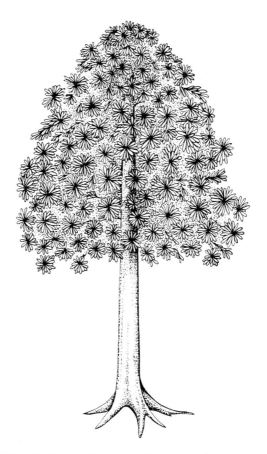

Text-figure 27. During the Permian, forests dominated by gymnosperms bearing *Glossopteris* leaves covered much of the southern middle and high latitudes. They were substantial trees that produced vast quantities of leaves, which have been preserved in the Permian rocks of places such as South America, southern Africa, India and Australia. Drawn by Annette Townsend, based on the work of R. E. Gould.

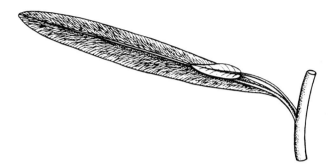

Text-figure 28. The reproductive structures of the glossopterid trees were typically borne on the leaves. This example shows a stalked, ovule bearing structure, known as a capitulum, of the genus *Lanceolatus*. A photograph of a fossil of such a fertile leaf is shown on Plate 97. Drawn by Annette Townsend.

attached to glossopterid leaves or shoots, suggesting that this was a very diverse group of plants – perhaps not a totally surprising fact, since they dominated the southern temperate forests for at least 40 million years (i.e. most of the Permian). Several ovules were attached to the expanded head of a sporophyll-like structure, which in turn was attached via a slender stalk to a leaf (an example is shown in Text-fig. 28). Although the intimate association between the foliage and the reproductive organs is not in doubt, there has been some disagreement as to whether these ovulate structures were attached to the midvein of the leaf or its petiole. The leaf may be the indistinguishable from normal, sterile leaves, or may be smaller depending on the genus. The head of the sporophyll-like structure may be flat or partly enrolled around the ovules, and in one case the laminate head entirely envelops a single ovule resembling a cupule such as seen in some other Palaeozoic gymnosperms. The male reproductive organs are less well known, but seem to be essentially similar to the female organs except that pollen sacs rather than ovules are attached to its head. Male and female organs seem to have been borne separately on different leaves, but there is no evidence that there were separate male and female plants, such as in modern *Ginkgo*. The idea that some of the fructifications were bisexual, which caused some palaeobotanists to speculate about the glossopterids being ancestral to the angiosperms, is now discounted as having been due to a misinterpretation of a complex female structure.

The glossopterid's nearest relatives are probably the caytonias, a group of Mesozoic gymnosperms, which are dealt with in the next chapter. However, their ancestors are a mystery. There were few gymnosperms growing at these middle latitudes during the Carboniferous; most of these areas were covered by glacial ice at that time, or supported only a very restricted tundra-like vegetation. The Carboniferous tropical vegetation had several groups of gymnosperm, but none had

Text-figure 29. Cordaites were trees and shrubs growing in the Late Carboniferous tropical swamp forests, and which were related to the primitive conifers, although the long, strap-like leaves were quite different in appearance. Drawn by Annette Townsend, based on the work of D. H. Scott, A. A. Cridland, G. W. Rothwell and S. Warner.

reproductive organs or foliage that are comparable. Recent discoveries in Antarctica and Australia of petrified glossopterid fossils, including reproductive organs, may help us understand better the detailed anatomy of these plants and to resolve some of the problems surrounding their ancestry.

CORDAITES

For the last group of gymnosperms to be dealt with in this chapter, we return to the tropical forests of the Late Carboniferous. They were trees or scrambling shrubs (Text-fig. 29) that occupied a range of different habitats, including mangroves and the drier areas surrounding the swamps. Unlike most of the contemporary gymnosperms, at least those growing in the tropics, cordaites did not have fern-like fronds, but instead had elongate, strap- or tongue-shaped leaves, with straight, parallel veins running along their length (Pl. 99). These leaves were helically and densely arranged on the stems, giving a very superficial resemblance to the modern *Dracaena* tree.

The smaller branches had a zone of conifer-like secondary wood surrounding a large central zone of pith. As these branches grew and elongated, no new pith was added and as a consequence gaps appeared in the central cavity, separated by thin septae of pith. As with the horsetails (see Chapter 4) this cavity often became filled with sediment after the death of the plant, resulting in a pith cast (known as *Artisia*) with numerous transverse lines marking the septae of pith.

The reproductive structures consist of cone-like fructifications (inflorescences) that were distributed among the leaves on the most distal branches. Each inflorescence consisted of two or four ranks of bracts attached along a slender axis, with a small cone attached in the axil of each bract (Text-fig. 30; Pl. 99). The cones themselves comprised helically arranged scales, most of which were sterile, but the most distal ones bore either an ovule or a number of pollen-sacs. The ovules were flattened, with two more or less prominent lateral wings. Pollen was very like that of the early conifers, in which an air-bladder (or saccus) surrounds the pollen grain to aid wind transportation. The inflorescences were either all male or all female, but it is not known if there were separate male and female plants.

From their similarities in stem anatomy and reproductive structures, there is no doubt that these cordaites were closely related to the early conifers, and they are usually included in the same class, the Pinopsida. It is tempting to take a progymnosperm ancestor such as *Archaeopteris* (see Chapter 2) and hypothetically to elongate its spirally attached leaves, to produce a cordaite-like plant. By subsequently reducing these leaves again, you could potentially get a conifer. Many palaeobotanists now think that this is much too simple a story, but there is nevertheless general consensus that the cordaites and conifers were closely-related sister-groups, perhaps both derived from a common ancestor among the archaeopteridian progymnosperms.

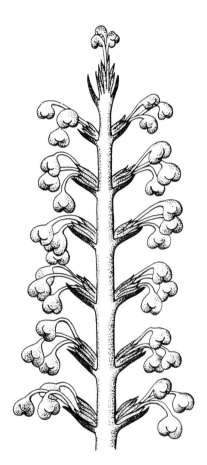

Text-figure 30. The reproductive structures borne by cordaite trees and shrubs were complex, consisting of a series of small cones attached to either side of a slender stem. The figured reconstruction is a female cone with ovules, similar to that shown as a fossil on Plate 99. Drawn by Annette Townsend.

The cordaite plants as outlined above are typical of the Late Carboniferous and Permian tropical forests, but very similar foliage occurs in the contemporary temperate floras. Especially in the northern temperate latitudes, plants producing such leaves (known as *Rufloria*) were very abundant and their remains have formed thick coal-deposits in Siberia. However, the associated reproductive structures are rather different, especially in the cones not being aggregated into inflorescences. Unfortunately, few fossils of these plants have been found with anatomy preserved, at least compared with the coal ball petrifications of the tropical cordaites, and so their affinities at the moment remain rather uncertain.

Chapter Seven

MODERN SEED PLANTS

The plants dealt with in this chapter are the gymnosperms that are typically found as fossils in the Mesozoic and Tertiary, a few of which have survived through to the present day. However, they are not sharply distinguished from the gymnosperms discussed in Chapter 6. Two of the groups discussed below have their first occurrences in the Palaeozoic, while at least one of the groups from the previous chapter ranges through into the lower Mesozoic (the peltasperms). Nor are there any specific characters that separate the two groups of gymnosperms. The division should be taken for what it is: a separation of a particularly large group of fossil plants into two parts, for the convenience of the authors and readers.

Included here are some of the best examples of 'living fossils' in the plant world. The maidenhair tree (*Ginkgo biloba*), for instance, represents a once widely-distributed group of plants that came to the verge of extinction, only to be rescued by horticulturists as an elegant garden tree. *Metasequoia* is a conifer that was first described as a fossil from the Tertiary and was only later discovered still growing in a remote part of central China. Again, horticulturalists have since planted this tree extensively in Europe and North America (some young examples can be seen along the roads near the Natural History Museum in London) thus rescuing it from probable extinction.

EARLY CONIFERS

Conifers are the most successful gymnosperms living today, consisting of over 600 species divided into seven or eight families. They have an almost worldwide distribution, but are especially common in the cooler high latitudes where they form extensive forests. They include the largest living organisms of any type: the giant redwoods of California (*Sequoia semperivens*). They also include some of the oldest living organisms, such as the bristlecone pines of Nevada (*Pinus aristata*) some of which are estimated to be nearly 5000 years old. Some conifers are small shrubs but most are substantial evergreen trees. Their foliage mostly consists of slender needle-like leaves attached helically to the branches, although some have broader leaves with several veins along their length (such as the Kauri pine

Agathis), and reproductive structures are separate male and female cones. Unlike angiosperm wood, conifer woods have no vessels running along their length.

Several types of conifer fossils are now known from the Palaeozoic. While they certainly have many resemblances to modern conifers, they are sometimes assigned to their own order (Volziales) due to differences in the construction of the ovulate cones. The oldest known examples are fragments of charcoalified conifer leaves found in middle Westphalian deposits in northern England. Although extremely small, they are beautifully preserved and show what seem to be typically conifer-like stomata. Towards the top of the Westphalian, more complete segments of conifer-like branches and, by the Stephanian, the oldest anatomically-preserved conifers from Oklahoma, Texas and Kansas can be found.

These early conifers are thought to have looked superficially like the modern Norfolk Island pine (*Araucaria excelsa*), having had rigid branches and scale-like leaves (Pls 100–101). Much of the interest in early conifers has focused on the structure of their cones, especially the female cones. Some of the earliest known cones in which some anatomical detail is preserved, from the Upper Carboniferous of North America (assigned to the family Utrechtiaceae), have a structure that invites comparison with their contemporaries, the cordaites. There are helically arranged bracts with a dwarf shoot in each axil. The dwarf shoot consists of several helically arranged sterile scales and one or more fertile scales. Each fertile scale bears one ovule attached laterally to the scale, with its micropyle facing the axis of the dwarf shoot (i.e. it is reflexed). This is unlike in the cordaites, where the ovule is terminally attached. (In the conifers, the terminology is different, so what was an inflorescence in the cordaites is referred to as a cone, and what were referred to as cones become dwarf shoots and ultimately ovuliferous scales.) Conifers and cordaites used to be thought of as directly related, but this view is not now so widely accepted because of the differences in attachment and orientation of the ovules. There are nevertheless so many points of similarity that it is difficult to imagine that they did not at least share a fairly close common ancestor.

In some other Late Palaeozoic conifers, the dwarf shoots have become somewhat flattened (family Emporiaceae) and in others the fertile scales are starting to become basally fused (family Majonicaceae). This trend continued with the Late Permian family the Ulmanniaceae, in which the scales have become fused into a single, lobed scale with ovules attached to some of the lobes.

The pollen cones of these plants have not been so widely studied. However, they appear to be simpler and more similar in structure than the ovulate cones, consisting of sporophylls arranged helically around a central axis, with two or more pollen sacs (number depending on the genus) attached to the stalk of each sporophyll.

These Palaeozoic conifers mostly occurred in palaeoequatorial vegetation. Conifer-like foliage can also be found in the southern, Gondwana floras, but the associated reproductive organs are quite different from those described earlier. The ovulate cones of the Ferugliocladaceae, best known from the Lower Permian of

Text-figure 31. The cheirolepidiacean conifers were an important member of the Mesozoic forests and had a distinctive, twisted shape to the trunk. This reconstruction of a large tree was redrawn by Annette Townsend from an original by Pauline Dean, and was based on the work of J. E. Francis.

South America, are simple cones without sterile bracts, while in the Permian Buriadiaceae from India, the ovules are not in cones at all, but are directly attached to the vegetative branches, replacing normal leaves. It seems that the presence of conifer-like foliage is mainly a reflection of the conditions to which the plants had become adapted (mainly drier habitats) rather than being unequivocal evidence of a close relationship. It is likely that the modern conifer families arose from one or more of the volzialean families of the palaeoequatorial belt, but there remains considerable uncertainty as to which. Where once there seemed to be a simple story of evolution from cordaites to volzialeans to modern conifers, there now seems to be a complex phylogenetic history, only part of which is directly preserved in the fossil record.

MODERN CONIFERS

As with the club mosses and many ferns, the most typically Palaeozoic conifer families become extinct at the P/T (Permian-Triassic) boundary. Only some of the Ullmanniaceae seem to extend through into the Triassic, and even they do not persist beyond the early Jurassic. Instead we see in the Triassic and early Jurassic the rise of many conifer families which are still alive today, such as the Araucariaceae, Podocarpaceae, Taxodiaceae and Taxaceae. (The latter family, which includes the yews, is sometimes regarded as not a true conifer partly because the ovules are not borne in cones.) Evidence of fructifications that are clearly indicative of these families has been found in the Jurassic, especially in the Middle Jurassic flora of Yorkshire (e.g. Pl. 104), and in some cases as far back as the Triassic. However, two of the most diverse of the living conifer families, the Pinaceae (including the pines and larches) and the Cupressaceae (including cypresses and junipers) do not have a reliable fossil record below the Cretaceous. Whether this is just because the fossil record is inadequate for these conifers (maybe they only grew in upland areas in the Jurassic), or it is because these two families are phylogenetically more recent, remains to be seen.

Much attention has been paid to the origins and evolution of modern conifer families. The research has revealed that, although some of the families can be traced back deep into the Mesozoic, there are also the remains of conifers showing a combination of characters of the modern families. For instance, the misleadingly named *Pararaucaria* from the Middle Jurassic Cerro Cuadrado flora of Patagonia is a petrified cone that combines features of the Taxodiaceae and Pinaceae. Eventually it may be hoped that evidence from such fossils will help unravel the complex evolutionary history of the conifers.

While most Mesozoic conifer families have survived to the present-day, there is one notable exception: the Cheirolepidiaceae (Text-fig. 31). This appears to have been one of the commonest conifers of the Mesozoic, with an almost worldwide distribution from the Late Triassic to Middle Cretaceous. The pollen (known as

Text-figure 32. The female cones of the cheirolepidiacean conifers released the seeds whilst still attached to the ovuliferous scale, in contrast to most modern conifers where the scale remains attached to the bract. This reconstruction shows a cone on the left still containing the seeds and scales, and a cone on the right that has shed them. Redrawn by Annette Townsend from an original by Pauline Dean, and based on the work of J. Watson.

Classopollis) in particular is highly distinctive, with a trilete mark, and a groove that extends around the grain about half way between its equator and the end furthest from the trilete mark. The cones had large, lobed ovuliferous scales, which were separate from the large bracts (Text-fig. 32). The scales were shed leaving cones consisting only of the helically arranged bracts, which are often found fossilized. Why the cheirolepidiaceans became extinct is still a mystery. It certainly had nothing to do with the K/T extinction event, as the youngest known fossils are middle Cretaceous. Perhaps it reflects increased competition from the more modern conifer families and the angiosperms, which seem to have been diversifying at about that time. There may also be an element of habitat-loss, as the middle Cretaceous was a time of rising sea levels, which may have flooded the flat coastal fringes where many cheirolepidiaceans were growing.

Before leaving the conifers, mention should briefly be made about their foliage. Conifer-like foliage is highly distinctive and has been widely recognized as fossils from the Upper Carboniferous to Recent deposits (e.g. Pls 100–101, 104, 105). However, distinguishing conifer families on foliage alone is less easy. If cuticles are preserved, this can often help, but many conifer fossils are merely impressions or poor compressions with no cuticles. To help overcome this problem, most palaeobotanists assign Mesozoic conifer foliage, unless reproductive structures are attached or there is other good evidence of their affinities, to one or other of about eight form-genera defined exclusively on the shape and attachment of the leaves. It has been shown in some of the better-preserved fossils, that foliage of more than one form-genus was attached to different parts of the same tree. It also means that some names that appear in the literature, such as *Cupressinocladus* (the form-genus for small, *Cupressus*-like leaves attached to the stem in decussate pairs or alternating whorls), cannot be used on their own as evidence of the family Cupressaceae. Nevertheless, provided the limitations of the scheme are kept in mind, these form-genera provide a useful way of recording conifer foliage found in the fossil record, without making any unwarranted claims as to their affinities.

GINKGOALEANS

Ginkgo biloba is probably the best example of a 'living fossil' among the plants. In the Mesozoic, the ginkgophytes were a highly diverse group with an almost global distribution; but today it is represented by just a single species. It was once thought to be totally extinct in the wild, the last remaining examples having been preserved in Japanese and Chinese temple gardens, where it was discovered by western botanists in the 1700s. However, a small stand of wild *Ginkgo* has recently been reported in Zhejiang Province, south-east China and there must remain the possibility that further stands may be discovered in some of the less well known parts of that large country. Why *Ginkgo* has undergone such a dramatic decline is far from certain. Because it is such an elegant tree with attractive foliage, it has been widely cultivated in Europe and North America, where it has proved to be a remarkably resilient and adaptable plant; hardly what one would expect of a species on the verge of extinction. It nevertheless seems that they suffered as a result of competition with angiosperms and became more and more restricted to northern temperate forests during the Tertiary. When ice spread over many parts of these temperate forests during the Pleistocene (the 'last ice age'), the main remaining habitats of the *Ginkgo* were destroyed.

Ginkgo-like foliage (e.g. Pl. 105) is known as far back as the late Triassic. There have even been putative ginkgophyte fossils reported from the Upper Palaeozoic, although most of these are now disputed. The problem with elucidating the evolutionary history of ginkgophytes is that relatively few reproductive structures are known. Some, such as *Allicospermum* from the Middle Jurassic of Yorkshire, are very similar to living *Ginkgo*. Ovules are borne at the ends of stalks (Text-fig. 33), usually in pairs, while the pollen is produced by loose catkin-like structures. However, some fossil ginkgophyte reproductive structures are very different from those of today's species and are assigned to the Ginkgoales mainly because of the associated foliage.

As with the conifers, identifying ginkgophyte foliage without reproductive structures is a problem (e.g. Pl. 105). Some palaeobotanists assign most ginkgo-

Text-figure 33. A typical short shoot of a *Ginkgo*-bearing both leaves and the paired ovules. Also shown is a fully mature seed. This reconstruction is based on the living *Ginkgo biloba*, but many of the Mesozoic forms would have been very similar. Redrawn by Annette Townsend from an original by Pauline Dean.

phyte fossil foliage to the modern genus *Ginkgo*, while others assign all such fossils to a separate form-genus *Ginkgoites* unless reproductive organs are known. The problem remains to be satisfactorily resolved, but a careful study of the morphological variability of the fossil leaves combined with cuticular evidence, should help improve the position.

Students of living plants have tended to regard *Ginkgo* as closely related to conifers. However, as evidence from the fossil record has progressively improved, possible links with the Palaeozoic cordaites or even the peltasperms have been suggested. Again, this remains an unsolved problem.

CYCADS

With eleven genera and some 185 species, this is one of the most diverse groups of living gymnosperms, second only to the conifers. They have a wide but rather unusual distribution, with most genera being restricted to particular geographical areas; no genus (with the possible exception of *Cycas*) occurs in more than one continent. It is now thought that these isolated genera are the remnants of much more widely distributed Mesozoic and early Tertiary populations, and indeed the fossil record indicates that the cycads were much more abundant and diverse, especially in the Mesozoic.

Modern cycads usually have an unbranched main stem, covered with a thick layer of persistent leaf-bases left after the shedding of old leaves. The stem may be long and thick, giving the plant a superifical resemblance to a palm-tree (e.g. *Macrozamia moorei*), short and squat (e.g. *Cycas revoluta*) or may be short and exclusively underground, like a tuber (e.g. some *Zamia* species). The leaves are large, compound fronds, which are usually rather thick and stiff. Reproductive structures are in the form of cones, with male and female structures being borne on separate plants.

One of the distinctive features of cycads is the way that vascular tissue enters the leaf from the stem. A strand of vascular tissue (a leaf trace) branches off from the stele of the stem but does not immediately enter the stalk (petiole) of a leaf. It instead divides into two strands, which pass around either side of the stem. These strands eventually enter the base of a leaf on the opposite side of the stem from where the leaf trace left the stele. This distinctive feature known as girdling leaf traces has been recognized in fossils as far back as the Triassic and seems to be unique to the cycads.

Cycads have many similarities with the medullosaleans dealt with in the previous chapter, and it is now generally accepted that they are descendants of that group. There are some fossils from the Upper Carboniferous that seem to show intermediate features between these two groups, strengthening the idea of this relationship. However, the oldest known unequivocal cycads come from the Lower Permian of China (Pl. 106). These include fronds and, most significantly, large

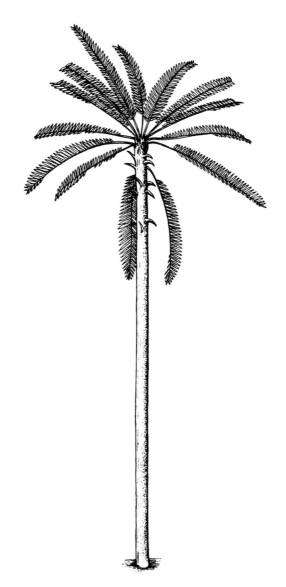

Text-figure 34. A Triassic cycad, *Leptocycas gracilis* Delevoryas and Hope, based on the work of T. Delevoryas and R. C. Hope (x 0.1). Although the leaves somewhat resemble the foliage of living cycads, the long, slender trunk is quite different. Drawn by Annette Townsend.

fimbriate sporophylls that are very similar to those seen in the cones of living *Cycas*. Other lines of evidence have suggested that *Cycas* represents the most primitive of the living genera of cycads, and this Chinese evidence seems to support this view.

The cycads reached their zenith during the Mesozoic, especially the Late Triassic to Early Cretaceous. Not all cycad-like frond fragments found in rocks of this age are true cycads (some belong to the bennettites dealt with in the next section). There are nevertheless abundant cycad remains in many floras of this age, including leaves (Pl. 107) with their distinctive cuticles and numerous fructifications. Some cones had fimbriate, *Cycas*-like sporophylls, similar to those seen in the Permian. However, there are also ovulate cones that are more reminiscent of the living Zamiaceae, in which the sporophyll has become significantly reduced to a slender stalk with a flattened head, the latter bearing the ovules. The relationship of these Mesozoic cycads to the living forms is not clear. The possibility that they had branching stems, and that the cones were much less rigid structures has caused some authors to place them in a separate family, the Nilssoniaceae. Others, however, regard them as merely early members of the Zamiaceae.

Although cycad fragments are relatively abundant in the Mesozoic, we are still a long way from confidently being able to reconstruct the whole plants. Published reconstructions have tended to make them look like modern cycads, with a stout, scaly trunk bearing an apical crown of fronds. There have also been reconstructions showing them with a long, slender trunk, somewhat resembling tree-ferns (Text-fig. 34). The suggested reconstruction of what has become known as the *Beania* tree (Pl. 108) shows the fronds being borne on a branching stem, although the author of this reconstruction (the eminent palaeobotanist Professor Tom Harris) confessed that the evidence for such a branched stem was not strong.

Fossil cycads are known through much of the Tertiary, but were clearly undergoing a significant decline from their Mesozoic heyday. As far as one can make out, this had little to do with the K/T event and was probably more a function of the cycads being out-competed by the more adaptable angiosperms and to a lesser extent by the conifers. Most cycads are extremely slow-growing plants, taking a long time to reach maturity. This can have its advantages in certain types of habitat, but has significant disadvantages when faced with competition from fast-growing and early-maturing plants such as the angiosperms.

BENNETTITES

Both cycad and bennettite remains occur abundantly in Mesozoic sediments, and share so many similarities that they can easily be confused. The bennettites had very similar shaped leaves to the cycads (Text-fig. 35; Pls 109–110) that are very difficult to distinguish on the basis of morphology alone. However, if cuticles are available, the two types of leaves can be told apart. The stomata on the lower

Text-figure 35. Leaves such as this occur commonly as fossils in many Mesozoic non-marine rocks. They may belong to either cycads or bennettites, and distinguishing these two groups of plants only on the form of the foliage can be very difficult. Either the epidermal structure of the leaves, or the form of the reproductive structures needs to be known before the foliage of the two groups of plant can be separated. These particular reconstructions are of leaves of the cycad *Nilsonia*, based on drawings by T. M. Harris (x 0.3). Drawn by Annette Townsend.

cuticle of cycad leaves consist only of a pair of guard cells surrounding the pore, whereas bennettites have a second pair of specialized cells (known as subsidiary cells) surrounding the guard cells (Pl. 111). The upper cuticle of bennettite leaves is also quite distinctive, being rather brittle and showing epidermal cells with very sinuous walls. These differences have proved consistent between these two groups of plants, allowing even quite small fragments of leaf to be separated.

The stems of some bennettites also resemble cycads, being squat and covered by a thick layer of persistent leaf-bases, and they superficially resemble a large pineapple in shape. Others have a long trunk and look more like one of the arborescent cycads such as *Macrozamia* (Text-fig. 36). Other bennettites, especially the earlier forms from the Triassic and Early Jurassic, had slender, branching stems without the layer of leaf-bases. Anatomically, there are again superficial similarities with cycad stems, with a thick central pith and a ring of secondary wood. Significantly, however, bennettite stems do not have the girdling leaf traces, that are such a characteristic feature of the cycads.

Despite the many apparent similarities between the bennettites and cycads, the reproductive structures show that their relationship was only distant. Some of the early bennettites had separate male and female organs. Most, however, had bisexual, flower-like reproductive structures (Pl. 112). These consisted of a central, dome-shaped body of tissue (similar to the receptacle of angiosperm flowers) on which were attached numerous small ovules separated by scales. The ovuliferous part was surrounded by a sheaf of pollen-bearing sporophylls, and these in turn were surrounded by a sheaf of protective bracts. In the earlier bisexual bennettite flowers, the pollen-bearing sporophylls were able to open out to facilitate wind dispersal. In the later species, however, the sporophylls were unable to open out. This undoubtedly provided protection to the pollen-producing bodies from predation, but also restricted the plants to self-pollination. This latter fact has been suggested as a major reason for the extinction of the bennettites in the Cretaceous.

The bennettites had an almost world-wide distribution in the Mesozoic. The earliest unequivocal examples are from the Upper Triassic of Austria. They are extremely abundant throughout the Jurassic and Lower Cretaceous, but decline in abundance and diversity in the Upper Cretaceous to disappear at the K/T boundary. This pattern of diversity suggests that the bennettites suffered at the expense of the angiosperm diversification towards the end of the Mesozoic.

These bisexual flowers caused many palaeobotanists to look at bennettites as possible ancestors of the angiosperms. It is now thought unlikely that there was a direct ancestor-descendent relationship, but rather that the bennettites are related to a rare living group of gymnosperms known as the Gnetales, which show an unusual combination of gymnospermous and angiospermous characters. Current thinking is that the bennettites, Gnetales and angiosperms probably arose from the same plexus of plants probably sometime in the very Early Mesozoic, or possibly even the Late Palaeozoic.

Text-figure 36. Bennettites were one of the commonest groups of plants in the Mesozoic. They had leaves very like the cycads, but the reproductive structures were very different. This reconstruction is of one of the Cretaceous bennettites, *Monanthesia* (x 0.05), and was drawn by Annette Townsend based on the work of J. Watson.

CAYTONIAS

Although the bennettites are the most widely distributed of the extinct gymnosperms groups in the Mesozoic, others have been found such as the *Caytonia*-plant (Text-fig. 37). Formally, the name *Caytonia* refers to female reproductive organs occurring widely in the Mesozoic Northern Hemisphere, but it has also become used to refer to the whole plant. They consist of two rows of ovule-bearing stalked cupules attached to either side of a stem. Each cupule has a more or less round outline, with an opening near its attachment to the stalk, and containing between 8 and 30 ovules, depending on the species (Pl. 113). It was originally thought that a lip at the opening of the cupule acted like stigma in angiosperm flowers, with pollen landing on the lip and entering the cupule via pollen tubes. However, it is now thought that pollination took place via a pollination drop, as with most gymnosperms.

Associated with *Caytonia* are clusters of elongate pollen-bearing structures (known as *Caytonanthus*) producing the same type of pollen as is found in the micropyles of the ovules. Also occuring are palmate clusters of three to six (usually four) leaflets at the end of a stalk. The leaflets, which have a prominent midvein and reticulate lateral veins, are very distinctive and can usually be identified from even relatively small fragments. It is from these leaf fossils, called *Sagenopteris*, that it has been possible to demonstrate the wide distribution of this group of plants in North America, Europe and central and eastern Asia, between the Late Triassic and Late Cretaceous.

The *Caytonia* plants attracted much attention in the earlier part of the twentieth century because they were though of as possible ancestors of the angiosperms. This was on the basis of the resemblance of their cupules to angiosperm ovaries, and the reticulate veining of the leaves. Both similarities are now recognized to be superficial, and the caytonias probably had little to do with the flowering plants. There is now some evidence that they may have been descendants of the glossopterids, which dominated much of the Permian vegetation in the Southern Hemisphere.

CORYSTOSPERMS

The corystosperms first appear in the fossil record in the uppermost Palaeozoic, but are best known from the Mesozoic. There are some similarities with the caytonias, to which they were probably related, but the differences are sufficiently marked that they are placed in a separate order: the Peltaspermales.

Like *Caytonia*, the ovules were borne in stalked cupules that were arranged in planated female structures. However, the corystosperm structures were much more complex, consisting of compound clusters of branches with cupules at the end. Also the cupules only contained one or two ovules. The pollen was produced by

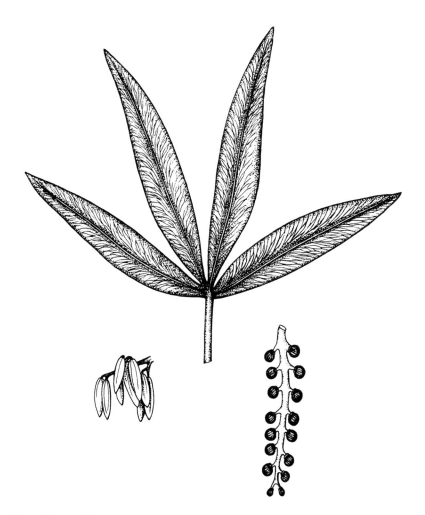

Text-figure 37. The *Caytonia*-plant had mesh-veined leaflets (centre), seeds enclosed in a carpel-like capsule (lower right), and clusters of pollen producing organs (lower left). Although such fossils have been widely found in Mesozoic rocks, we still do not know the overall form of the plant that produced them; whether it was a tree, a shrub, a vine or a herb. Drawn by Annette Townsend at natural size, based on the work of T. M. Harris.

elongate sporangia, but these were attached in clusters to a flattened sporophyll rather than individually at the end of an axis.

Corystosperm foliage had an underlying pinnate form, but was otherwise very variable. In Europe, the commonest type of leaf is *Pachypteris*, which was an exclusively pinnate frond with slender, tongue-shaped pinnules (Pl. 115). They could vary from once- to three-times pinnate. The leaves from the Southern Hemisphere, in contrast, tend to have a major fork near the base of the frond, dividing it into two pinnate segments, the best known example being *Dicroidium* (Pl. 114)

From the evidence of the foliage, the corystosperms seem to have had an almost world-wide distribution. However, it was in the Southern Hemisphere that they were most abundant and where they dominated many floras, especially in the Triassic. They generally became much rarer in the Late Cretaceous, eventually becoming extinct at the end of the Cretaceous. As with the bennettites and caytonias, it is difficult to get direct evidence of this decline in the Late Cretaceous: did they succumb to the environmental impact of the K/T boundary event, or were they already in significant decline in response to competition from the rapidly diversifying angiosperms? What is certain is that none of these Mesozoic groups of gymnosperms have any descendants left in today's vegetation.

CHAPTER EIGHT

FLOWERING PLANTS

Flowering plants, or angiosperms as they are more properly called, have dominated most terrestrial ecosystems of the world for the last 100 million years. Today, there are nearly 240,000 species in a bewildering variety of forms. They live in a greater range of environments, show a greater range of growth habits and morphological variation, and are represented by more families, genera and species than any other group of living vascular plants. Angiosperms form the basic diet of most herbivorous animals today and we, ourselves, rely on them for agriculture, horticulture and many pharmaceutical products. Their importance to man has inevitably caused much interest in the origins and early evolution of angiosperms, and there has been considerable work done on their fossil history.

WHAT MAKES AN ANGIOSPERM?

Defining what is an angiosperm is very difficult. The concept of a plant with showy and obvious flowers is not tenable for many angiosperms, so how might such plants be identified if found fragmentary in the fossil record? The range of characters recognized in living angiosperms is of course large in such an enormous group with so much plasticity of form. There are some features, which together can be taken to prove angiosperm affinity:
- the presence of a flower;
- enclosed ovules or seeds;
- a double protective layer – integuments – around the embryo sac;
- wood with vessels;
- the development of the food conducting phloem;
- multi-layered and tectate pollen walls consisting of pillar-like structures called collumellae, supporting an outer covering called the tectum;
- reticulate venation pattern in the leaves; and
- the distinctive double fertilization mechanism where two male nuclei fuse with nuclei in the female egg cell. (The presence of the double fertilization mechanism is often given as the main argument for their monophyletic origin, but proponents argue that such an evolutionary event could quite possibly have occurred more than once in 100 million years.)

In practice, vessels in wood, the reticulate venation of leaves, and pollen are the most commonly used diagnostic features of fossils, even though none are exclusively found in angiosperm and not all angiosperms have them. The gymnosperms *Gnetum*, *Ephedra* and *Welwitschia* have vessels, while some angiosperms such as *Tetracentron* (Hamamelideae) and some of the Winteraceae (Magnoliales) have none. *Gnetum* also has laminated leaves with reticulate venation, as do some ferns (e.g. *Hausmania*) and gymnosperms (e.g. *Sagenopteris*) while many angiosperms do not. Some angiosperms have non-tectate pollen, while some conifers, notably the Cheirolepidiaceae, are known to have tectate pollen. Flowers and some seeds, although much rarer as fossils, are the most distinctive organs. They also tell us more about the family affinities than can the other fossilized organs, as the classification system of extant angiosperms is largely based on floral characteristics.

ANCESTORS OF THE ANGIOSPERMS

For many years the origin of the flowering plants was thought to be a complete mystery, and ideas of their ancestors were based upon anatomical comparisons with living and fossil gymnosperms. However, in order to progress, the origin and evolution of the angiosperms must be treated as a problem waiting to be solved, rather than as a mystery, and to do this we must look at the clues.

Flowering plants, like any complex large plant, are preserved as bits and pieces in the fossil record. Of these, leaves and pollen grains are the most common, with anatomically preserved stems and then flowers being much less frequently found. A major problem faced by palaeobotanists is how to be certain that such isolated fossil organs are unequivocally angiosperm in origin, and seeking features which might be taken as indicators of ancestral angiosperms or angiosperm precursors is even more of a problem. More than twenty Mesozoic taxa have been considered seriously to be possible ancestors of the angiosperms, which shows how difficult recognition can be. The evolution of the angiosperm flower is itself even a matter for debate. There is no clear transitional series between gymnosperm reproductive structures and the angiosperm flower, so a number of complex hypotheses have been developed to explain how it could have happened. These are usually based on possible homologies between the position and morphology of organs making up the reproductive structures, but as yet the matter has not been satisfactorily resolved. Suggestions for ancestral groups have included the Gnetales, glossopterids, caytonias, cycads, bennettites and even ferns.

The angiosperms must obviously be descendants of a group that first evolved the critical combination of angiosperms characters discussed in the previous section. These characters then gave a competitive edge to the plants, enabling them to out compete their less angiosperm-like contemporaries. Evolution from the earliest Mesozoic groups could be taken to imply that the flowering plant line was distinct by the late Triassic and some biochemical evidence suggests that its origin was even earlier. In contrast, fossil evidence based on adpressions and pollen suggests that both their appearance and their main evolutionary radiation were in the Cretaceous. Proponents of earlier dates have postulated that angiosperms evolved in upland areas that were far from sites of possible fossilization. However, a subsequent widespread and successful invasion of closed gymnosperm-dominated lowland floras by such cryptic upland plants does not seem a very likely scenario. It should be made clear that we are not certain even whether the angiosperms are really a single group or have their origin in more than one evolutionary event. If the 'angiosperm threshold' was passed simultaneously by more than one group, subsequent evolutionary change and convergence would have obscured the original differences.

THE EARLIEST ANGIOSPERMS

It is generally accepted that the earliest flowering plants in the Cretaceous were most probably perennial herbs or small shrubs that lived in the more open and disturbed habits alongside rivers. This was a time of global warmth; the poles probably free of ice and the lower latitudes in the Early Cretaceous dry or seasonally dry. The Early Cretaceous vegetation was divided into several provinces that were dominated by different assemblages of gymnosperms. The climatic conditions of the time may have stimulated the evolution of such weedy plants with their photosynthetically efficient planated leaves, reticulate leaf venation (that provided an efficient transport system even when damaged), better vascular systems, and flowers that enabled a quicker and more efficient reproduction. It is interesting that the Gnetales also arose at this time, because they show many parallels with the flowering plants having flower-like reproductive organs, vessels in the wood, and large veined leaves in *Gnetum*.

The first magnolia-like leaves are rare, even when they are at their most abundant in gymnosperm and fern dominated fossil plant assemblages. This suggests that these plants grew in the forest as a subordinate element, a role that some plants retain to this day.

THE RISE OF THE MONOCOTYLEDONS

The origin of the monocotyledons is another problem that confronts us. Is such a disparate group that includes the grasses, palms and lilies really a monophyletic group and did they have a separate origin from the dicotyledons? We do not know. The earliest unequivocal monocotyledons appeared in the Late Cretaceous. There is palm wood (Pl. 116) from the Coniacian and fronds from the Maastrichtian. Grass pollen may possibly have been found in the Maastrichtian, but the earliest generally accepted records come from the Early Tertiary (Palaeocene) of Brazil, Africa and Australia. The earliest fossilized grass leaves are also known from the late Palaeocene.

THE IMPACT OF ANIMALS ON ANGIOSPERM EVOLUTION

Many early flowering plants must have been an attractive source of food with their leaves, flowers and fruits being more easily digestible than conifer shoots and cones. They were small and easily accessible to browsing, but could withstand its effects by being able to regenerate themselves, better than any of the gymnosperms. This must have been especially important for young saplings, which would certainly have given the flowering plants a great competitive advantage over all the conifers.

Interestingly, the mid-Cretaceous diversification of angiosperms is accompanied by the transition from a sauropod-dominated dinosaur fauna to one dominated by the browsing ornithopods that had efficient grinding teeth. The larger amount of food supply provided by the angiosperms led to the herbivorous dinosaurs increasing in number and diversity. The largest flesh eaters, such as *Tyrannosaurus*, then evolved to prey on them, resulting in a food chain similar to that seen today with mammals.

There is some evidence of leaf herbivory as far back as the Carboniferous, and by the Jurassic there is clear evidence of leaf miner damage in the gymnosperm *Pachypteris*. The flowering plants with their broad, comparatively fleshy leaves, were, however, a much more attractive source of food than conifers. Within 25 million years of early flowering plant radiation, insect herbivores were exploiting them as their major food source. Flowering plants were able to survive this constant onslaught through their ability to shed their older leaves and replace them with new ones.

Some flowering plant seeds can survive the digestive tracts of the animal browsers, which greatly aids their dispersal. The evolution of succulent seed coats and the swelling of ovaries to give fruits added to their attractiveness as food and improved their potential dispersal.

The other interesting plant/animal interaction that the rise of the flowering plants produced was pollination by insects. For effective reproduction, pollen must germinate on the receptive surfaces of the female reproductive organs. The means by which pollen is transferred is varied, but in living angiosperms insects are among the commonest vectors, with complex protein recognition processes usually ensuring that only pollen of the right type germinates to effect nuclear fusion. Although insect pollination occurred in some gymnosperm groups, it was developed to its highest level in the angiosperms. Such plant-pollination relationships reward the pollinator for visiting the plant while they incidentally effect

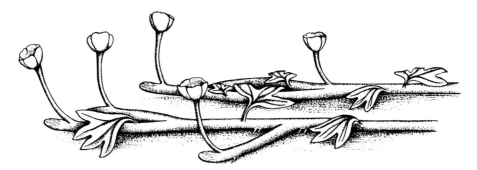

Text-figure 38. A reconstruction of the Early Cretaceous (Wealden) angiosperm, *Bevhalstia* (x 1.5). It was probably an aquatic plant or grew on mud flats. Drawn by Annette Townsend, based on the work of C. R. Hill and E. Jarzembowski.

pollination. The earliest reward was the pollen itself, but the development of nectar together with more organized and specialized flower parts led to many different co-evolutionary paths of flowers and their pollinators.

WIND POLLINATED ANGIOSPERMS

Not all flowers were insect pollinated and there is a view that non-biological pollination, particularly by wind, was secondarily derived. Wind-pollinated flowers are typically unisexual and they may, or may not, be on the same plant. This may not be consistent in families and even within a genus. The male flowers are often arranged into catkins and the earliest well-documented catkin-like structure is known from the Cenomanian of Kansas, USA. There are also ball-like seed heads known from roughly the same locality that are very similar to those of extant *Platanus* (planes). There is even leaf and pollen evidence of the pollen-producing mimosids from Kansas, but the first evidence of inflorescences is from the Eocene. Leaf evidence suggests that the birches evolved in the Cretaceous but again reproductive evidence is not known until the Oligocene.

CRETACEOUS ANGIOSPERMS

Original ideas of the early evolution and radiation of flowering plants were almost entirely derived from studies of Cretaceous fossil leaves. The leaves were closely compared with those of extant families and often referred to extant genera. Newer studies not only show that practically all of the earlier identifications were

  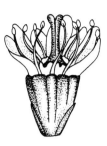

Text-figure 39. Reconstructions of the flowers of a Late Cretaceous angiosperm, *Silvianthecum* (x 12), based on charcoalified fossils such as shown in Plate 124. These small simple flowers indicate affinities with the Saxifragaceae. Drawn by Annette Townsend, based on the work of E.-M. Friis.

incorrect, but that successively younger Cretaceous flowering plant floras show the increasing levels of complexity that are predicted by many modern classification systems. The earliest (Early Cretaceous) leaves from central Asia, the Russian Far East, Portugal and eastern USA were small, simple, entire-margined and pinnately-veined and some were irregularly lobed. Middle Cretaceous leaves were quite varied, either pinnately compound or palmately lobed with more regular palmate venation patterns. Others had a low level of vein organization, while one magnolia-like form had the same vein pattern seen in the living members of the Magnoliales. Such close similarities must be treated with caution for they must not be taken as unequivocal evidence of a close taxonomic relationship. There were more or less circular leaves with radiating major veins that probably came from climbers or possibly aquatics. Another major group had palmate veins and sometimes toothed margins, typified by the leaves of the *Araliosoides*, *Sassafras*, and the *Platanus* types.

For many years, it was assumed that flowers were too delicate to have become fossilized, but an increasing number of flowers have recently been found, many of which are referable to extant families. The first convincing flower fossils are Late Cretaceous in age (e.g. Text-fig. 38; Pl. 124), although Early Cretaceous fragments may subsequently be proved to be of flowers (Text-fig. 39). Pollen grains provide an additional means for monitoring change and evolutionary radiation through time. The oldest known pollen grains, from the Lower Cretaceous, were simple and of the type restricted to magnoliids and monocotyledonous plants. Soon after this, four other classes of flowering plants were recognized by their pollen. Recent finds of macrofossils support the idea that the first recognizable flowering plants had small and simple flowers of the *Magnolia*-type, with no differentiation of petals and their surrounding green sepals (e.g. Text-fig. 40; Pl. 117). The pollen produc-

Text-figure 40. A fruiting axis of a primitive magnolia-like angiosperm, *Lesqueria*, from the Middle Cretaceous (x 1). The helically arranged carpels are very similar to those of the living *Magnolia*. Drawn by Annette Townsend, based on the work of P. R. Crane and D. L. Dilcher.

ing anthers and the ovule-bearing carpels were simple and relatively few in number.

A considerable amount of information now exists about the evolution and diversification of angiosperms throughout the Mesozoic and the effect that they had on the vegetation as a whole. Evolutionary change was certainly rapid because, by the Cenomanian, a number of distinct families had appeared. Apart from the magnoliid forms, there were now relatively primitive hamamelids (witch hazels), laurels, planes and simple rosids with buttercup-like flowers. Other well-differentiated flowers in the fossil record also provide evidence of a number of 'experimental' families that are now extinct.

By the Middle Cretaceous, angiosperms were starting to dominate some parts of the vegetation. There were freshwater swamp woodlands dominated by plants bearing *Magnoliophyllum*, *Liriophyllum* and *Sapindopsis* leaves, platanoid-leaved plants were abundant on channel levees and around lakes, and there were mangrove-like communities on the tidal deltas.

The time was clearly right for the flowering plants, because there was now a massive evolutionary burst, producing many other families before the end of the Cretaceous. The Maastrichian floral radiation was greater than in any previous or subsequent period. It can be related to changing climate and plate tectonics, which led to tropical conditions with rain forests that were unparalleled since the early Permian. Many flowers were now more specialized and some, like saxifrages, had well-developed nectaries to attract hymenoptera and stingless honeybees. The low level of specialization in floral morphology, coupled with indiscriminate pollen transfer and the probability of genetic compatibility between such early members of radiating groups, probably led to high levels of interbreeding. This would have accelerated morphological variability and new character combinations that might have enabled the hybrids to out-compete their parental populations possibly leading to population segregation. Isolation could then lead to further genetic and morphological adaptation resulting in better defined species boundaries.

By the end of the Cretaceous 31 extant families of flowering plants are clearly recognisable in the fossil record including the Magnoliaceae (magnolias), Hamamelidaceae (witch hazels), Juglandaceae (walnuts), Caryophyllaceae (pinks), Leguminosae (peas) and the Araliaceae (ivies). Angiosperms abscise their older leaves as they grow and a further adaptation to periodical drought during their early evolutionary radiation could have permitted some species to shed all their leaves synchronously to precede a dormant phase. This deciduous habit would have permitted angiosperms to colonize areas subjected to periodic water or temperature stress where the gymnosperms could not survive. It also permitted them to spread to higher altitudes and latitudes where lower winter light and temperatures could be survived in a dormant condition. There is evidence that many angiosperms migrated northwards during the late Cretaceous and this was most probably along the coastal plain. Once north of the closed conifer forests they would have spread away from the coast into the interior.

TERTIARY ANGIOSPERMS

Angiosperms appear to have been little affected by the Cretaceous-Tertiary (K/T) extinction event (see Chapter 9) that caused so much havoc with animal life. During the Palaeogene and Neogene periods, angiosperms continued to diversify rapidly, eventually dominating most of the world's land vegetation. Our knowledge of angiosperm evolutionary history is still far from complete, with only about a half of the 300–400 living families having been identified in the fossil record. Of these, however, just under a hundred first evolved during the Palaeogene (see Appendix 1). Especially in the Northern Hemisphere, the Ulmaceae (Pl. 118),

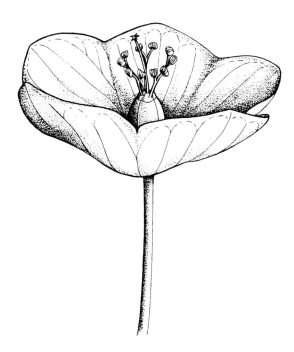

Text-figure 41. Reconstruction of a *Florissantia* flower from the Early Tertiary, based on fossils such as that shown in Plate 121. These plants are thought to have been related to the extant Malvales. Drawn by Annette Townsend, based on the work of S. R. Manchester.

Betulaceae (Pl. 119), Juglandaceae (Pl. 120) and Salicaceae (Pl. 122) became major elements in the flora. A major factor in explaining this Palaeogene explosion in diversity is probably the global climatic cooling that started in the Middle Eocene. This would have increased the stress on vegetation, which had become largely adapted to the warmer conditions of the Mesozoic and early Palaeogene, and would have provided an opportunity for diversification in the more adaptable groups of plants, such as the angiosperms.

This climate change continued through the Neogene, culminating eventually in the Quaternary ice age. The most important Neogene development in land vegetation was the expansion of open grasslands. Although grass fossils are known from the early to middle Palaeogene, the pollen record suggests that the grasslands proliferated during the early Neogene (Miocene), especially in the low and middle latitudes. This vegetational change has had a profound impact on animal life, providing the trigger for the evolution of the herbivorous ungulates that have become among the most numerous mammal groups. It also provided the ancestral forms of the cereal and forage crops on which man is now so dependent.

As in the Mesozoic, much of our knowledge of Tertiary angiosperms is derived from leaf fossils (e.g. Pl. 119). Angiosperm foliage studied in isolation from the rest of the plant can be very difficult to interpret, but considerable progress has been made in recent years through the detailed analysis of venation patterns and leaf-margin characters. Some flowers have been found (e.g. Text-fig. 41; Pl. 121) but they are generally rare. More abundant are fossil fruits and seeds, and these tend to be a reliable means of identifying angiosperm families (Pl. 123). There are a number of classic Palaeogene fruit and seed floras, such as from the London Clay of southern England and the Clarno Nut Bed of Oregon, and these have provided a clear insight into the rapid diversification that was taking place in angiosperms during the Palaeogene.

If isolated organs such as fruits and seeds are examined, they often appear to be very similar to those of their living descendants. In some cases, this has resulted in Palaeogene fossils having been assigned to living genera, suggesting that relatively little evolutionary change had taken place over the intervening 50 million or so years. In recent years, though, it has been realised that a proper understanding of angiosperm evolution can only be obtained by studying at least partly reconstructed plants. From this, it is evident that, although individual plant organs may seem to have changed little, the Palaeogene angiosperm plants would often show quite a different combination of organs. For instance, one species from the Palaeogene of southern England had a winged nut similar to that of the living hornbeam, while the arrangement of the bracts is more similar to that of the hazel, and the foliage is different again from that found on either plant today. It is becoming clear that considerable evolutionary change has taken place among the angiosperms during the Tertiary; a fact that should not come as a total surprise in view of their evident adaptability and success in today's vegetation.

Chapter Nine

THE HISTORY OF LAND VEGETATION

Through most of this book, we have looked at the fossil record of the major groups of plants. In this chapter, we will attempt to stand back and take a broader look at the plant fossil record, to examine how land vegetation as a whole has changed through time. When and where did the major developments in the evolutionary history of plants take place, and what were the major types of vegetation at any particular time?

TIME AND STRATIGRAPHY

Although time is an integral part of the study of palaeontology, absolute time (i.e. how old something is in years) still presents some significant problems. The sequence of events is relatively easy to establish by looking at their relative occurrences in the sequence of rocks. If event 'A' can be observed in rocks that overlie the rocks in which event 'B' is recorded, it is reasonable to assume that event 'A' occurred after event 'B'. This is one of the underlying principles of the science of stratigraphy: the study of time as revealed in the succession of rocks.

Until relatively recently, getting any idea of absolute ages in the fossil record was little more than guesswork. Attempts to estimate time by working out rates of sedimentation of muds and sands and relating that to the thickness rocks produced highly inconsistent results. The major breakthrough was the development of radiometric dating, which estimates the age of rocks by the relative proportions of isotopic breakdown products. This has recently been supplemented by other methods such as fission-track dating such that we can now relate the fossil record to a fairly broad-brush chronology; we know roughly how many years ago plants moved on to the land, when the first seeds evolved, and how old the earliest known angiosperms are. The problem remains, however, that the number of dates that have been established in the geological record are relatively few and they all have quite wide ranges of error. This means that for day-to-day purposes, palaeontologists still usually use the traditional stratigraphical nomenclature for describing the chronological sequences of events in the fossil record.

This is not the place to give a detailed account of stratigraphical nomenclature and methodology (see the section on further reading towards the end of this book). Suffice it to say that the sequence of rocks and fossils can be divided up according to their relative chronological positions (chronostratigraphy). These divisions are classified using a nested hierarchy of time units called eons, eras, periods, epochs and ages, each of which is assigned a name. The accompanying figure shows the eras and periods of the Phanerozoic, which is the eon that concerns this book (the underlying Proterozoic and Archaean eons have only a very restricted fossil record, and appear to predate the appearance of land vegetation). The following account will summarize the vegetational characteristics of each of the periods from the Silurian to present, giving their absolute dates in millions of years (MA).

SILURIAN PERIOD (410–437 MA)

Land vascular plants first evolved in the Middle Silurian (Wenlock Epoch), their earliest representatives being small rhyniophytes or rhyniophytoids and club-mosses (Chapters 2 and 3).

DEVONIAN PERIOD (355–410 MA)

From its primitive rhyniophyte and lycophyte precursors, land vegetation rapidly diversified during the Devonian. By the end of the period all of the major divisions of vascular plants except the flowering plants had appeared: ferns (or at least their pre-fern ancestors) by the Middle Devonian, and horsetails and gymnosperms by the Late Devonian.

The Devonian also saw the development of several structures that made the vascular plants better adapted to life on land. Secondary wood is first seen in the fossil record in the Middle Devonian, this being the most widely adopted means by which plants can significantly increase their stature. The resulting trees (and therefore presumably forests) will have had a dramatic impact on the Devonian landscape. Photosynthetic efficiency was also enhanced by the development of planated leaves in the Devonian. The process seems to have been gradual through the Devonian and probably developed independently in the horsetails, ferns and progymnosperms.

The main reproductive novelty to appear in vascular plants during the Devonian was the seed. Although this was again a gradual evolutionary process via heterospory (see Chapter 2), the earliest structures that can properly be called seeds are found in the Upper Devonian. The seed freed plants from the necessity of having external moisture to achieve fertilization and thus allowed it to occupy a much wider range of habitats. It also provided the young plant with a source of nutrition, giving them the edge over homosporous plants, which had to generate nutrition (usually) by photosynthesis much sooner after germination. Finally, seeds were able to remain dormant for a period of time between fertilization and germination, allowing them to take advantage of favourable conditions whenever they occurred.

It is difficult to assess vegetational diversity and provincialism in the Devonian. Most of our knowledge comes from Europe and North America; our knowledge from elsewhere is still quite poor. The traditional view was that global provincialism was relatively low, especially in the early and late Devonian, with some provincialism recognizable in the Middle Devonian. However, new work in places like China and South America is revealing a much greater geographical diversity in the Devonian than previously realized, but we are still a long way from establishing even the broad patterns in the distribution.

CARBONIFEROUS PERIOD (300–355 MA)

The morphological developments that took place in plants in the Devonian provided the springboard from which vegetation could start to take full advantage of the terrestrial habitats during the Carboniferous. There were few new morphological developments in the Carboniferous. Planated foliage became more common and fructifications more complex (e.g. the development of cones), but these were in essence just refinements of the structures first seen in the Devonian. The conifers and cycads (or at least cycad ancestors) make their first appearance, but neither of these 'modern' groups is widely represented in the Carboniferous fossil record. What we do see, however, is the spread of plants over large parts of the earth to produce a vegetation pattern that is in many ways similar to today's (although, of course, the individual plants were very different).

In the Early Carboniferous, vegetation in middle and high latitudes was relatively poor, dominated by club mosses and progymnosperms. In low latitudes, such as Europe and North America, a much more diverse vegetation was flourishing; as well as there being a greater diversity of club mosses and progymnosperms, there were seed-ferns (lagenostomaleans and calamopityaleans), pre-ferns and true ferns, and horsetails (archaeocalamites). Much of the evidence reflects marginal lowland habitats such as lake margins and river deltas. However, there is also some record of plants growing in volcanic terrains, in which the mineral rich waters have often petrified the plants, preserving part of their internal anatomy; sites in southern Scotland have produced many examples of this type of preservation.

The Late Carboniferous saw a dramatic change in vegetation, coincident with the start of the Permo-Carboniferous ice age. Much of the southern low-latitude land was covered by thick ice and as a consequence there was little or no vegetation. High northern latitudes were mainly deep ocean, but there is some evidence of sea-ice having formed at that time. The northern middle latitudes contained relatively impoverished floras dominated by horsetails and primitive pteridosperms.

In low latitudes, however, things were very different, with the appearance of the very first tropical rain forests: the so-called Coal Measures forests of Europe, North America and China, named because of the vast reserves of coal-forming peat that were laid down. These forests were dominated by the giant club mosses (Chapter 3), which became perfectly adapted to the wetland habitats that then existed over much of the then tropics. Also within the forests, mainly on somewhat more raised, drier ground (e.g. river levees) a more diverse range of ferns (including marattialean tree ferns), horsetails (calamites and sphenophylls), seed ferns (medullosans, callistophytes, lagenostomaleans) and cordaites. Surrounding these wetland forests were probably conifer-dominated forests, although the remains of these habitats only occasionally found their way into the fossil record.

Towards the end of the Carboniferous, this tropical wetland forest mostly disappeared, except for the Far East (mainly China) and a few localized pockets in Europe. This was probably due to the effects of mountain building that was then going on over much of Europe and North America (known as the Variscan Orogeny) that caused the area to become drained. The giant club mosses that dominated the forests were so tightly adapted to the wetland conditions that, when the habitats were drained, they simply died out. At the very end of the Carboniferous, the coal-forming forests briefly returned to North America, but this time dominated by tree ferns and seed-ferns rather than club mosses. However, this resurrection was short lived and the forests disappeared again in the Early Permian.

PERMIAN (250–300 MA)

This period marks the culmination of the first great development of vegetation on land. Several groups of plants, which made their first tentative appearance in the fossil record in the Carboniferous suddenly, become significantly more abundant, including conifers, cycads, glossopterids, gigantopterids and peltasperms. Seed plants became the dominant trees, supplanting the club-mosses, horsetails and tree-ferns as the commonest large land plants. Ferns and horsetails became mainly represented by herbaceous plants; some tree-ferns persisted but they never returned to the abundance they achieved in the very late Carboniferous. Some pockets of swamp habitat with giant club mosses continued through the Permian in the tropics of China but, here again, these plants never returned to the abundance achieved in the Carboniferous.

In the western tropics, in North America and Europe, the Variscan mountain building that started in the Late Carboniferous saw the rapid disappearance of the coal-forming swamp forests. In Europe and easternmost North America, where the tectonic activity was at its maximum, the aridification resulted in vegetation that was largely dominated by conifers. The only evidence of significantly more diverse vegetation was on the margins of the large inland Zechstein Sea during the

Late Permian, where there is evidence of horsetails, peltasperms and cycadophytes, as well as conifers.

Further west, such as in Kansas, Oklahoma and Texas, there is evidence of a more diverse vegetation, including gigantopterids, peltasperms and conifers. This is strikingly similar to the Permian vegetation of large parts of China, which is also dominated by gigantopterids, peltasperms, cycads and conifers. This may in effect be regarded as the 'typical' palaeoequatorial lowland vegetation of the Permian, which was only prevented from spreading into Europe because of the tectonically-induced aridification.

In higher latitudes, vegetation at the start of the Permian was limited due to the persistence of polar glaciation. However, when this glaciation collapsed during the Early Permian, abundant vegetation developed in the higher latitudes of both the northern and southern hemispheres. This vegetation resulted in thick deposits of peat, that have produced some of the worlds major coal reserves, such as those of Siberia, Kazakhstan, South Africa, India and Australia. Extensive forests of glossopterids characterised the vegetation of the southern higher latitudes, while cordaites and ruflorias dominated the northern forests. As the polar ice declined and eventually disappeared during the Permian, a marked increase in global vegetational provincialism can be recognized. Although plants could now grow in much higher latitudes, there were nevertheless restriction on what types of vegetation could survive the conditions of extreme seasonality. In the very high southern latitudes of what is today Antarctica, the forests were dominated almost exclusively by the glossopterids, but in somewhat higher latitudes, such as India, South Africa, South America and Australia, there were also horsetails, peltasperms, cycads and conifers. Similarly, in the very high northern latitudes of what is today northeastern Siberia, the forests were largely dominated by the cordaite-like plants, while the middle latitudes, such as seen in central and western Siberia, Mongolia and Kazakhstan did more diverse forests flourish.

PERMIAN-TRIASSIC EXTINCTION EVENT

The end of the Permian saw the most dramatic change ever to occur to the world's vegetation. The tropical swamp forests disappeared with the extinction of the giant club mosses, and the higher latitudes lost the cordaites and glossopterids. It was all part of the massive Permian-Triassic extinction event, which caused some 96 per cent. of the known species of both animal and plant to become extinct. Among the gymnosperms, for instance, of the nineteen families known from the Permian only three range through into the Triassic. Its effects can be seen in the fossil record throughout the world and clearly represented a cataclysmic event that came close to exterminating all life on the planet.

Despite the immensity of the biotic change that occurred at this time, we are still very much in the dark as to its causes. Despite searches for evidence of asteroid or comet impact, such as the iridium layer at the Cretaceous-Tertiary boundary (see below), an extra-terrestrial explanation seems unlikely. Large areas of flood-basalts in Siberia and pyroclastic deposits in South China dating from about this time have caused some geologists to suggest that extensive volcanism caused atmospheric pollution and climate change. The most recent suggestion has been that a sudden change in oceanic circulation at the end of the Permian resulted in stagnant deep-oceanic waters, rich in carbon dioxide and hydrogen sulphide, being brought up to near the surface. The resulting surge in these gases into the atmosphere (known as the 'big belch') would have had the combined effect of global climatic warming and depleted oxygen levels in the atmosphere, which together may have been sufficient to have had such a dramatic effect on the Earth's biota.

Whatever the underlying cause, this greatest of all mass extinctions for land vegetation represents the change from the primitive Palaeozoic floras to floras of a recognizably modern aspect that start in the following Triassic period.

TRIASSIC PERIOD (205–250 MA)

Immediately following this extinction event, land vegetation was significantly depleted. This is reflected in an almost complete absence of coal deposits of this age anywhere in the world. The fossil record reveals that the Early Triassic lowland vegetation was dominated by the club moss *Pleuromeia*. During the Middle and Late Triassic, however, many modern-day families of ferns and conifers started to appear, together with several now extinct groups, such as the bennettites. These groups probably first evolved in the Palaeozoic or very Early Mesozoic in upland habitats, which are not normally represented in the fossil record, eventually taking advantage of the lowland habitats when they were vacated by the mass-extinction at the end of the Palaeozoic. The result was the typical Mesozoic vegetation that dominated the lowland habitats between the Middle Triassic and Late Cretaceous.

The most species-rich Late Triassic floras were in equatorial latitudes, such as Germany and south-western USA, where a range of ferns, horsetails, pteridosperms, cycads, bennettites, leptostrobaleans, ginkgos and conifers occur. The general balance of the flora within the equatorial belt tended to be uniform; what differences there are between, for instance, North America and China was mainly at the species level. Middle latitudes had generally similar floras, albeit not as species-rich. Bennettites and dipteridacean ferns were absent from northern middle latitudes (Siberia and Canada), while in the southern middle latitudes (southern Africa, Australia) the cheirolepidiacean conifers are unknown. On the whole, however, it seems that there was little latitudinal variation in the lowland Late Triassic vegetation, probably reflecting a fairly uniform, frost-free climate over much of the world.

JURASSIC PERIOD (135–205 MA)

As in the Triassic, the low latitudes had the most diverse vegetation in the Early and Middle Jurassic (western North America, Europe, central Asia and the Far East), which included ferns, horsetails, pteridosperms, cycads, bennettites, leptostrobaleans, ginkgos and conifers. By now, most ferns belonged to families that are still living today (e.g. Dipteridaceae, Matoniaceae, Gleicheniaceae, Cyatheaceae). The conifers of this age can also be mostly assigned to modern families (e.g. Podocarpaceae, Araucariaceae, Pinaceae, and Taxaceae), although one of the most abundant families of this time became extinct at the end of the Cretaceous (the Cheirolepidiaceae). Coals are often associated with this type of vegetation, reflecting high rainfall and lush vegetation.

The exception to this pattern was in the western part of the tropical belt (central and eastern North America, and North Africa), where desert conditions prevailed in the Early to Middle Jurassic. During the Late Jurassic, this arid area extended throughout much of the low latitudes. The rare plant fossils from this arid area reflect a mainly impoverished vegetation dominated by cheirolepidiacean conifers and matoniacean ferns (*Weichselia*). The desert did not extend as far as northern Europe, but conditions here nevertheless became much drier and the dominant plants (bennettites, cycads, peltasperms, cheirolepidiacean conifers) tended to have thick cuticles with sunken stomata, features normally associated with dry conditions. Coals were rarely formed here.

A vegetation-type known as 'Ginkgoalean taiga', mostly dominated by ginkgos and leptostrobaleans, characterized the northern middle latitudes (Siberia and north-western Canada). Extensive coal deposits were formed here, suggesting conditions were relatively warm and humid. This is supported by the virtual absence of cheirolepidiacean conifers. Cycads and bennettites only became relatively common here in the Late Jurassic, especially in eastern Siberia.

The vegetation of the southern latitudes in the Early to Middle Jurassic was similar but less diverse than that of the equatorial belt, except for the scarcity of plants such as the ginkgos. In contrast to the northern latitudes, there were abundant cheirolepidiacean conifers. There was little or no polar ice at this time and so this southern vegetation extended into very high latitudes, including Antarctica.

During the Late Jurassic, the equatorial desert extended into parts of the southern middle latitudes (South America and Africa), but elsewhere there was a diverse vegetation of this age, often including horsetails, ferns, cycads, bennettites, ginkgos and conifers. One of the most famous fossil floras of this age is from the Rajmahal Hills in India, which has yielded beautifully preserved petrifactions showing anatomical structure.

CRETACEOUS PERIOD (65–135 MA)

The vegetation of the Early Cretaceous differs little from that of the Late Jurassic, both in distribution and broad composition. Low latitudes were dominated by arid, desert or sub-desert conditions (South America, central and North Africa, central Asia), where the vegetation was mainly of cheirolepidiacean conifers and matoniacean ferns. Northern middle latitudes, such as in Europe and North America, contained a more diverse vegetation including bennettites, cycads, ferns, peltasperms and cheirolepidiaceans. Further north, the diversity again declined, to be replaced by leptostrobalean dominated assemblages. The southern middle latitude floras appear to be dominated by bennettites and cheirolepidiaceans.

However, the Late Cretaceous saw the start of a major change in style of vegetation, marked mainly by the proliferation of the angiosperms (see Chapter 8), with a corresponding decline in some of the typical Mesozoic elements, such as leptostrobaleans, bennettites, ginkgos and cycads. It in effect marks the beginning of the end for the typical gymnosperm-dominated Mesozoic flora, and the onset of the development of today's angiosperm-dominated vegetation.

During the Late Cretaceous, arid conditions continued to extend over most of the low latitudes. What little is known of the tropical vegetation of this time (from northern South America, central Africa and India) was dominated by palms and proteas, with few if any conifers. Palms also extended in to northern middle latitudes. The most diverse floras of this time were in North America, which were dominated by evergreen angiosperms, with some conifers especially the redwood *Sequoia*.

The absence of polar ice meant that angiosperms extended into high latitudes at this time, and their remains have been found very near to the then north pole. In contrast to the trees of the middle latitudes, however, these high latitude angiosperms were not suprisingly deciduous. Conifers were also more common at these higher latitudes and were increasingly common further north, especially the taxodiaceans (*Sequoia*, *Metasequoia*). Interestingly, relicts of more typical Mesozoic vegetation still occurred in parts of these high latitudes such as Siberia, with elements such as caytonias, bennettites and leptostrobaleans.

We know relatively little about the vegetation of the southern middle latitudes in the Late Cretaceous, which were largely covered by the same desert that affected the low latitudes. What is available suggests that horsetails, ferns, conifers and angiosperms dominated the vegetation fringing the desert. The commonest conifers were araucarians and podocarps, while the angiosperms include *Nothofagus* (southern beech), all of which are typical Southern Hemisphere genera of today. As in the Northern Hemisphere, conifer and broad-leaved angiosperm forests seem to have extended into very high southern latitudes, suggesting there was little if any polar ice.

CRETACEOUS-TERTIARY EXTINCTION EVENT

In several places in the world, the boundary between the Cretaceous and Tertiary (K/T) is marked by a very distinctive layer of clay with unusually high levels of iridium. This high level of iridium, which is normally a very rare element on the surface of the Earth, is thought to be the result of either an asteroid impact probably near Mexico, or massive volcanic activity in India (or possibly both). Whatever the cause, it seems that the end of the Cretaceous was marked by a period of darkness and lowered temperatures.

It is now well-known that the dinosaurs suffered extinction at this time, as did some marine animals such as the cephalopods and some planktonic foraminifera. However, the event had less of an effect on plant life. Very few families became extinct at this time, and those that did (e.g. the bennettites and caytonias) were already suffering significant decline, probably as a result of competition with the angiosperms. The darkness and lower temperatures would undoubtedly have had an effect on plants, killing both their vegetative and reproductive organs. The flowering plants that had never been exposed to low temperatures would have been particularly susceptible. However, the seeds and spores that had already been released from the plants, and which were lying dormant on or in the ground, would mostly not have been effected by the environmental disruption and would have been capable of germination when conditions became more normal.

The temporary effect of this sudden 'ecological shock' can most readily be seen in the fossil record from the middle palaeolatitudes of present day western North America. Immediately above the K/T boundary there is a marked dominance of ferns in the pollen and spore record. This is analogous to the early colonization of land laid bare by volcanic activity such as in Krakatau (Far East) in 1883 and Mount El Chichon (Mexico) in 1982. The subsequent Tertiary successional colonization in North America showed a selection in favour of deciduous taxa and an elimination of the evergreen vegetation that was dominant before the trauma. This catastrophe has been interpreted as a short freeze of possibly less than a year, rather than a longer impact winter, because it did not affect the warmer and more southerly floras, nor the more northerly ones that were already adapted for deciduousness. The effect certainly appears to have been relatively local because there is no comparable vegetation trauma documented elsewhere. Because there is no evidence for a mass kill other than in the middle latitudes in North America, it was certainly not a global event.

PALAEOGENE AND NEOGENE PERIODS (1.6–65 MA)

The first vegetation types characteristic of the modern world appeared in the succeeding Tertiary. Many Tertiary plant fossils are identified closely with living taxa and often included within their systematic groupings such that extant genera are commonplace. The fossil record of Tertiary plants has also become linked to the concept of ecological comparisons and the migration or extinction of plants due to climatic change. This is, of course, too simple because plants can also respond through evolutionary change. Latitudinal difference in floras was apparent in the early Tertiary. There was a distinct flora growing around the Arctic Ocean. This was a broad-leaved, deciduous and stratified forest of angiosperm trees, shrubs and herbs, within which extant families diversified. In North America, the masskill of the vegetation at the K/T boundary was followed by a period of increased rainfall, making the conditions right for the widespread development of multistratal rainforests in the more southerly areas. This development is identified in the fossil leaf floras by an increase in leaf size, the high percentage of species with apical drip-tips (a feature of many of today's tropical rainforest plants) and vines. Northwards there were temperate broad-leaved evergreen forests and mixed conifer forests. Finally there was a broad-leaved deciduous forest, the polar Arcto-Tertiary forest, found on the north-west Canadian archipelago at a palaeolatitude of 75°–80° north, where mild moist summers at about 25° alternated with continuous winter darkness at about 0°–4°C. This primary deciduous vegetation included members of the Platanaceae, Judlandaceae, Betulaceae, Menispermaceae, Cercidophyllaceae, Ulmaceae, Fagaceae and Magnoliaceae with gymnosperms belonging to the Taxodiaceae, Cupressaceae, Pinaceae and the Ginkgoaceae. This flora was able to spread between Europe and North America until the land bridge between them was severed in the Palaeocene. Similar interchange between North America and Asia was possible by the land bridge in the Bering Straits from the Late Cretaceous to the Late Neogene. In North America a mid continental sea way separated the European integration from that with Asia. Even now, the floras of the south east have more common elements with Europe than with those of western North America.

A different more southern flora was growing around the Tethys Sea, the southern USA and Asia. This lower latitude flora has been described as of a tropical type because many plants have close affinities with those growing in today's tropics. Much of the information comes from a study of seed and fruit floras found in Europe and North America (e.g. Pl. 123). The largest and most varied of these is the London Clay flora which includes about 350 species. The interpretation of the vegetation is based on an assessment of the nearest living relatives and their ecological tolerances. Although some of the ecological tolerances may have changed over the last 50 million years, this approach does provide some indication of the palaeoclimate. Many of the London Clay species have their nearest living relatives among the tropical vegetation of the Indo-Asian region, but there are many others whose affinities are with living temperate species. The London Clay flora, like other Tertiary floras, cannot be precisely equated to any flora or climatic condition that exists today. The closest analogue is a semitropical, or pantropical, rainforest which contains all the elements of the tropical rainforest except the taller dipterocarp trees that stand above the main forest canopy. Today such floras are delimited by the 20° and 25°C mean annual temperature isotherms in eastern Asia, being replaced by tropical rain forest at the upper limit. The dense forest had abundant shrubs and climbers with the plants of 'temperate affinity' growing along the stream sides and in open habitats within the forest. The shoreline and the banks of the larger rivers would have been covered with mangrove which was dominated over large areas by the stemless-palm *Nipa*. Reed-swamp with plants like extant *Scirpus* bordered the rivers above the reach of tidal influence.

However, within two million years many of these 'tropical' elements had disappeared from Europe, having migrated southwards. At this time, the Arcto-Tertiary flora was fragmenting and it migrated southwards during the Eocene/Oligocene cooling and the onset of the Pleistocene fluctuations. For the first time in the history of the angiosperms, there were large areas of temperate deciduous forests. In Europe waves of Arcto-Tertiary elements such as *Liquidamber, Ulmus, Alnus, Populus* and *Acer* penetrated the primary evergreen forests that were defined by laurophyllous elements. This was more pronounced during the middle and late Eocene, when there were mixed forests with much higher numbers of deciduous summer-green species. The best surviving examples of the European earlier Tertiary floras are now confined to the wetter parts of the Canary Islands.

At the end of the Tertiary, in the Miocene, there was a marked change in the vegetation in many areas. Open habitats increased in which grasses and other herbaceous plants flourished. This was the first time that prairie vegetation developed where herds of herbivorous mammals roamed.

QUATERNARY PERIOD (1.6 MA TO PRESENT)

During the Quaternary Period, spanning the last two and a half million years or so, there were dramatic changes in climate. It alternated between times of extreme cold, called glacials or ice ages, when much of north west Europe, Siberia and northern North America were covered with ice, and times of relative warmth, called interglacials, when the climate was similar to, or sometimes warmer than, today. There are fossil wood, leaves and fruits of some of the species (e.g. Pls 126–128), but our knowledge of Quaternary floras mostly comes from studies of pollen grains and spores recovered from peat, soil, or lake sediments. Samples prepared from cores taken from these sediments show the relative abundance of different types of pollen and spores that indicate the type of vegetation existing at the time. Pollen assemblages taken from different depths in a core can be compared to reveal how the vegetation in that area changed over time. A comparison of cores from a widespread area shows the vegetational changes over the whole region. These studies reveal that the Quaternary climatic oscillations resulted in large-scale migrations of both plant and animal species southwards and northwards in response to cooling and glacial advance and warming and glacial retreat.

The final migration of plant species after the last ice age, which reached its maximum extent 18–20,000 years ago, led to the present geographical distribution of new vegetational assemblages throughout the world. Many northern areas now have significantly fewer species than once lived there, especially where new seas surround isolated islands, such as the British Isles, that were once part of a continental land mass. The migration also led to some unusual distributions of species. For example there are arctic-alpines whose name reflects their disjunct arctic and continental mountain top distributions. This separation came about with climatic amelioration when some plants migrated northwards while others moved higher up the mountains where they eventually became marooned.

The study of post glacial fossil pollen and spores even allows us to follow the course of vegetational change brought about through human interference. The most favourable climate was about 7000 years ago when tree cover was at its maximum extent spreading much further northwards and higher into the mountains. Climatic deterioration lead to some retreat but by then the effects of man's selective woodland clearance for agriculture was taking effect. Such clearance continued around much of the world with the obvious effects that we can see today. That is, however, outside the scope of this book.

CHAPTER TEN

HIGHLIGHTS OF PALAEOBOTANICAL STUDY

Plant fossils have been studied for about three hundred years and right now there are several hundred active palaeobotanical researchers around the world. It is, therefore, not possible to give a full account of the history of palaeobotany in just a few pages. Instead, we give some highlights of the subject and concentrate on those publications, events and people that have influenced and developed the subject. It is, of course, a personal view and readers who are interested in knowing more are recommended to read the general texts in Andrews and in Lyons, Wagner and Zodrow and the other references listed for this chapter.

THE BEGINNINGS OF PALAEOBOTANY

Investigations of plant fossils really started in the late seventeenth and early eighteenth centuries when people such as Edward Llwyd (1660–1709), Martin Lister (1638–1711), John Ray (1627–1701), John Woodward (1665–1722) and Johann Scheuchzer were collecting and illustrating them. Through their publications they stimulated thought on whether these fossils were really the remains of once living organisms or just mineral accidents. The general reasoning came down on the side of once living organisms, although religious belief at the time constrained ideas and committed people to believe that fossils were formed as a result of the biblical flood. Nevertheless, there was a general acceptance that the plant fossils could be recognized as species even though it was thought that they could all be related to living species.

Progress in palaeobotany was slow during the rest of the eighteenth century with only a few publications of any note. It was still the time of the amateur naturalist and such men as James Parkinson (1755–1824) and William Martin (1767–1810) were adding to the general knowledge of the range of plant fossils that could be found. James Parkinson published his *Organic Remains of a Former World* in three volumes between 1804 and 1811. He clearly understood that coal was derived from plants and that fossils were important for stratigraphical correlation, but could not separate either notion from his belief in the biblical flood. Parkinson was in fact a physician by profession and was the first person to recognize the disease that is now known by his name. At the same time there was increasing knowledge of the natural world as explorers came back from around the world with herbaria and living plants for the new botanic gardens. The subject eventually came to the fore in the early part of the eighteenth century with the publication of a number of important works that were to place the study of plant fossils on a secure scientific footing. Three names dominate the start of this era of palaeobotany: Ernst von Schlotheim (1764–1821), Kaspar Maria, Graf von Sternberg (1761–1837) and Adolphe Brongniart (1801–1876).

Schlotheim was the first and produced many palaeobotanical works, though his most important was *Die Petrefactenkunde*, not published until 1820. Schlotheim had collected throughout Germany and France and clearly understood the stratigraphical significance of his fossils. He also realized that Carboniferous plant fossils could not be closely compared with living species and put forward the idea that they had lived in a warmer climate than was found in the fossil locality today, a rather bold idea for the time. Unfortunately, although Schlotheim's work has been accepted as scientific, it is Sternberg's publication, *Versuch einer geognostich-botanischen Darstellung der Flora der Vorwelt* [An attempt at a geographical-botanical description of prehistoric times], that has been taken as the official start for naming plant fossils. Published in parts, between 1820 and 1832, it is a comprehensive and well illustrated account of fossil plants from what Sternberg believed to be the three periods of vegetation: the periods of coal plants, cycads and flowering plants – effectively the Palaeozoic, Mesozoic and Tertiary periods as accepted today. The third person was Brongniart. He and Sternberg were both interested in living plants and their knowledge of them influenced their palaeobotanical studies. This is perhaps most obvious in Brongniart's greatest work, the *Histoire des Végétaux Fossiles*, in which he classified his fossils with a system used for living plants. The work was published in parts between 1828 and 1838, but ended abruptly, in mid sentence, before he dealt with the seed plants.

An interesting point to consider here is the length of time over which these major works were published. As each part came out it reflected both an increased knowledge derived from new discoveries and the influence of ideas from other publications. Two important English publications appeared during this time frame. Edmund Artis's *Antidiluvian Phytology* appeared in 1825 and included comparisons with fossils figured by Schlotheim, Sternberg and Brongniart, remarks on the systems used by these authors and comments received from the famous geologist William Buckland (1784–1856). Artis even named one of his fossils *Sternbergia* after Sternberg, and Sternberg reciprocated with *Artisia*. More important was the *Fossil Flora of Great Britain* by John Lindley (1799–1865) and William Hutton (1797–1860), which was published between 1831 and 1837 and contained 230 plates and descriptions of plant fossils from the Carboniferous to the Tertiary. Although, like the *Antidiluvian Phytology*, it was more of an illustrated catalogue than an authoritative text it did provide others with good illustrations on which to base their ideas and theories.

Back on the continent, Heinrich Göppert (1800–1884) was the next important palaeobotanist to arrive on the scene and he started publishing on plant fossils in 1835. Then in 1836 his important *Systema Filicum Fossilium* appeared, which included a 76-page history of palaeobotanical endeavours so far. Göppert published widely and in a thesis (1838) figured pollen that he had recovered from a fossil alder catkin. Unfortunately it was to be many years before the significance of recovering fossil pollen and spores was to be realized. Soon after this, in 1884, Wilhelm Schimper (1808–1880) of Strasbourg published on the Triassic plants of the Vosges. Then, after working on living bryophytes for many years, he returned to palaeobotany with his monumental *Traité Paléontologie Végétale*, published in three volumes together with an atlas of plates between 1869 and 1874. It was conceived as a complete manual of systematic palaeobotany with what a contemporary described as a highly scientific and rational classification system.

Palaeobotany was on a much surer footing in the latter part of the nineteenth century, but one must remember some of the constraints on thinking that prevailed at the time. Religion was still a very powerful influence and many people thought Darwin to be a heretic when in 1859 he published his ideas about evolution in his *Origin of Species*. The other major constraint lay in the acceptance of the stability of the world as it is today. The German climatologist and meteorologist Alfred Wegener (1880–1930) first published the idea of continental drift in 1915, but it did not receive much attention until an English translation appeared in 1924. Even then Wegener's idea was very controversial and argument raged for many years. Until continental drift became accepted, the ideas of land bridges and cataclysmic events were used to explain the distribution patterns of both plants and animals.

THE INFLUENCE OF COAL

The latter part of the nineteenth century saw momentous changes with the industrial revolution sweeping through Europe. The new industries were powered by coal so it is not surprising that great effort was put into discovering new coalfields and working all the profitable seams within them. The fossil plants found in the roofing shales of many of the coal seams quickly passed from being mere curiosities to providing a possible means of comparing the ages of different coal fields and the stratigraphic positions of their various coal seams. Palaeobotany therefore attracted many able people who were often associated with coal mining, surveying or transport. Eventually such comparisons were worked out not only within Europe, but also between Europe and North America.

The Frenchman Francois-Cyrille Grand'Eury (1838–1917) started life as a mining engineer, but later became Professor of Trigonometry at St-Étienne. One of his greatest achievements was to assemble whole plants from their disparate parts. This he did through careful observation and judging the evidence cautiously before committing himself. His reconstruction of the cordaite tree is still used today.

Grand'Eury was, however, convinced of the need to use distinct generic names for the different organs, even if he was convinced that in this instance they were parts of the same original plant. Both of his principles are used today.

Another Frenchman, René Zeiller (1847–1915), was one of the first people to use plant fossils seriously in Carboniferous and Permian stratigraphy, publishing his three large memoirs in 1884, 1888 and 1906, which are still of use today. Zeiller was very interested in Carboniferous ferns and also described the male fructifications of pteridosperms for the first time.

The Scot Robert Kidston (1853–1924) was a contemporary of Zeiller although, unlike him, Kidston did not have to work for a living. At the age of twenty-six Kidston became financially secure and devoted the rest of his life to palaeobotany. His main interest was in the Coal Measures and he published many accounts of plant fossils from British coalfields. This eventually led to his immensely important monograph on Coal Measure plants. Six parts were published between 1923 and 1924 in which were figured the very best of the superb collection that he had gathered during his many visits to the coalfields. Unfortunately Kidston died in 1924 while visiting a colliery agent and amateur palaeobotanist, David Davies (1871–1931), in South Wales before completing his monograph; see the accounts by Thomas and Edwards for further details. The remaining six parts were completed by Robert Crookall (1890–1981) and published between 1955 and 1970, some even after Crookall himself had retired, but they were too late to have any major impact on the development of palaeobotany. If they had been published in the 1920s, as intended, they would have contributed greatly to the Coal Measures stratigraphical work that was being done at the time. However, in the intervening years Carboniferous palaeobotany had made considerable advances, especially through the work of continental palaeobotanists, and the Crookall monographs did not take this into account.

People like Emily Dix (1904–1973) and later Leslie Moore were actively working on the stratigraphy of South Wales. In her pioneering work, in the 1930s, Dix distinguished floral zones in the South Wales coalfield which she later attempted to use on the continent. Unfortunately Dix suffered a mental breakdown in the late summer of 1945, from which she never recovered. The most notable of the other researchers at the time were Walther Gothan (1879–1954) who was working in Germany, Wilhelmus Jongmans (1878–1957) in The Netherlands, Paul Bertrand (1879–1944) in France, Armand Renier (1876–?) in Belgium and William Culp Darrah (1909–1989) in the USA. Through their own work, their cooperation in exchanging ideas and indeed fossil plants, and the occasional disagreement they each extended their knowledge of Carboniferous plants and stratigraphy beyond their original country commitments, and devised floral sequences to determine the comparative ages of Coal Measures sequences.

This work should be set against the changing situation in Europe. The 1914–1918 war had finished and Europe was at peace again, although conditions were of course different in the east because of the revolution in Russia. International co-operation was made easier with more modern transport and people were travelling greater distances. So in 1927 there was the first of the congresses on Carboniferous stratigraphy in Heerlen. These important congresses have continued with the last, the thirteenth, being held in Krakow in 1995. Stratigraphical research, of course, continues and there is much still to be learned. In North America, Charles Read (1907–1979) and Sergius Mamay developed an Upper Palaeozoic floral zonation scheme that has survived with little modification for over thirty years. In Europe, Robert Wagner, a student of Jongmans, while working in Spain recognized in the late 1970s a new Stephanian stage between the top of the Westphalian and the base of the Stephanian as it was then recognized. This new stage he called the Cantabrian.

The emphasis was not all on stratigraphy, for many palaeobotanists preferred to study their fossils botanically. New techniques were evolved to help this. John Walton (1895–1971) devised a method for turning fossils over: he stuck them face-down to glass slides and dissolved the rock away with acid. This technique often reveals the reproductive organs on the underside of leaves, which were left embedded in the rock when the fissure plane of the rock splitting ran along the smooth side of the leaf.

The emphasis of all the work discussed so far has been on adpression material, which of course gives the collector an instant picture of the plant remains. However, the anatomy of plants has mostly to be determined from petrifactions, which tend to be more difficult to study. The first studies on such material were undertaken by two Englishmen, Henry Whitam and William Nicol, both of whom published accounts of 'petrified wood' in the early 1830s. They used the pioneering technique of grinding thin, transparent sections that were mounted on glass slides for microscopic study. This work stimulated others to spend their time and energy researching the internal anatomy of fossil plants. Edward Binney (1812–1881) was one who did just that, but more important was his 1885 report, with the great botanist of the time Joseph Hooker (1871–1911), on the structure of plants found in coal balls. These carbonate concretions found in certain coal seams opened up a whole new field for anatomical study that has yielded so many immensely important discoveries of botanical and evolutionary significance. Coal-ball study was first taken up in an enthusiastic way by William Crawford Williamson (1816–1895), whose palaeobotanical career had commenced with drawing the Yorkshire Jurassic specimens figured in Lindley and Hutton's *Fossil Flora* (referred to earlier). Williamson's work included nineteen important memoirs based largely on coal balls, and perhaps the most memorable is his study on the lepidodendroid rooting base *Stigmaria* (see Chapter 3). The large specimen of *Stigmaria* that he removed from a Pennine quarry at his own expense is still there to be seen in the Manchester Museum. Very soon after this time, other *Stigmaria* rooting bases were found. In 1873 a grove of ten was discovered during excavations for a mental asylum at Wadsley in Sheffield, although unfortunately only four now remain. Another grove was discovered in a Glasgow quarry in 1887, but these were preserved in a covered enclosure and remain safe today. Williamson collaborated later in life with Dunkfield Henry Scott (1854–1934), who went on to become one of the leading palaeobotanists of his time. Williamson's 1896 autobiography makes interesting reading on his life, the times and the people with whom he worked.

Coal balls were first recorded from the USA in the late 1890s, although these were pyritic and only able to be studied using reflected light. In the 1920s, carbonate coal balls were discovered in the Illinois coalfield. Adolphe Noé (1873–1939) was the first to undertake palaeobotanical investigations in the early 1920s, aided by his students, most notably J. Hobart Hoskins (1896–1957). William Darrah (1909–1989), Aureal Cross, Henry Andrews, James Schopf (1911–1978) and Wilson Stewart carried the work forward. John Walton re-enters the story here by devising, in 1923, the peel technique referred to in the introductory chapter. Walton used cellulose compounds that took a long time to dry and it was not until much later that William (Bill) Lacey (1923–1995), with co-workers, introduced the idea of using acetate sheets. Lacey's technique was a great leap forward in coal-ball study because it permitted serial sections to be made through the smallest of plant organs with the minimum of waste. There are many more coal-ball localities in the USA than in Europe so most of the mid and late twentieth-century work on them has been American. The main exception has been the work on coal balls from the Donetz basin in the Ukraine. From this work has come a much greater understanding of Carboniferous plant morphology, reproduction and ecology.

The peel technique is not limited to coal balls. Albert Long also used it to great effect on seeds and stems found in Lower Carboniferous permineralisations (see his 1997 autobiography). His early work was on coal balls in Lang's laboratory in Manchester, but after a period of amateur entomology while school teaching, he turned his attention to Lower Carboniferous palaeobotany. In the Scottish borders, where he then lived, there had been some earlier work on petrified material by William Gordon (1884–1950), Professor of Mineralogy at Kings College, London. Long turned his attention to this field of study and found beautifully preserved material that had never been described. In a series of great papers he described many simple pteridosperm seeds and even found some of them still in their rudimentary cupules. Long also turned to the study of the tree *Pitus*, which had first been described by Whitham in 1833, and showed it to be a pteridosperm with fern-like foliage, seeds in cupules and microsporangia.

The study of coal itself became an important issue because of the differing qualities, and therefore value, of the coal recovered from different seams. One of the pioneers in the study of coal was Marie Stopes (1880–1958) who, with Professor Wheeler, published in 1918 an account of the constitution of coal in which they originated the terms clarain, durain, fusain and vitrain for the different petrographic types of coal. A modified form of this classification is still in use today. Marie Stopes had first researched palaeobotany in London and Manchester and travelled to visit other palaeobotanists, even visiting Japan, before she became interested in coal. Then in 1918 she also published her controversial book *Married Love* and in 1921 opened the first scientific birth control clinic in the world. She had thrown herself headlong into controversy and from then on she had little time for plant fossils, although she did continue to attend scientific meetings for the rest of her life. Marie Stopes pioneered a social revolution, while writing books, plays, poems and political pamphlets. She was certainly the most colourful palaeobotanist of all time.

It was also of great importance to be able to use coal samples to identify coal seams discovered in prospecting boreholes, especially in areas of severe underground faulting. From this need emerged the study of isolated spores prepared from the coal and from the sediments associated with them. Henry Witham is accepted as being the first person to figure spores seen in a thin section of coal. P. F. Reinsch published the first attempt at an artificial classification of microspores isolated from coal in 1884, while James Bennie and Robert Kidston in 1886 gave the first excellent descriptions of megaspores. Further work was, however, slow to come because there appeared to be no interest or value in it. A number of Europeans did, however, start to publish accounts of dispersed spores in the 1930s and the literature on this topic gradually grew. A. C. Ibrahim in 1933 published a morphological classification of pteridophyte spores and used a binomial system of nomenclature that is still basically in use today. One other pioneer was Arthur Raistrick, who published a number of papers between 1932 and 1940. His 1934 paper on correlating coal seams in Northumberland was met with great enthusiasm and hailed as a novel method with high promise for the future.

The first monographic works to try to bring order to this rapidly expanding subject came from the Illinois Geological Survey in the USA. Schopf, Wilson and Benthall in 1944 and Kosanke in 1950 put the relatively new subject of palynology firmly on a scientific footing and their publications instantly became the standard reference works. Palynologists around the world started to specialize either on microspores or megaspores. The Dutch led the way in Europe in megaspore studies, with Sijben Jan Dijkstra (1906–1982), who worked at Heerlen with Jongmans. Dijkstra published a great deal on Carboniferous megaspores from 1946 onwards and much of it was on foreign material, either brought back by his boss Jongmans or sent by Jongmans' overseas contacts. The Germans Robert Potonié (1889–1974) and Gerhard Kremp started publishing in 1955 a series of monographs on spores from the Ruhr coalfields, which again became standard works. However, megaspores have never really proved to be very good stratigraphic markers, possibly because of the tendency of researchers to have very broad species and to lump other authors' species into their own.

Microspores have proved to be much better stratigraphical markers and have been used by many more palynologists. Interestingly there is not the same tendency to lump species in microspore research. Indeed there is often the tendency to divide species or create new ones. A landmark publication came in 1967 from the British National Coal Board with Harold Smith and Mavis Butterworth's work on microspores from British coal seams. Incidentally, this was the first of the splendid series of special publications by the Palaeontological Association. From its use in Coal Measure stratigraphy the subject of palynology became an essential tool in other exploration work, most notably in that of the petroleum industry.

As the number of dispersed spore species increased many researchers attempted to relate them to their parent plants. This was made possible by recovering spores from fructifications and comparing them to the published accounts of dispersed spores. The first time such *in situ* 'spores' were recorded was in Heinrich Göppert's *Les Genre des plantes fossile*, published between 1841 and 1846. Göppert (1800–1884) figured pollen recovered from a well-preserved alder catkin. However, apart from this early account there were only spasmodic records of *in situ* work until the early part of this century. The surge of palynological work in the late 1940s to 1960s naturally spurred on *in situ* studies from both adpressions and coal-ball permineralizations. Robert Potonié compiled the available information in his in *Synopsis der Sporae in situ* (Synopsis of *in situ* spores), the first part of which was published in 1962. From a knowledge of the parent plants came the possibility of using assemblages of dispersed spores for floristic and ecological interpretation. In the early 1960s Harold Smith set about interpreting Carboniferous palaeoecology on dispersed spores data. In the same year Harold Smith, working for the British Coal Board, illustrated a changing spore assemblage that could be correlated with coal petrography essentially using Marie Stopes' four coal-types. This he interpreted as reflecting a changing plant assemblage from a closed lepidodendoid-dominated swamp to an open moorland-type of community dominated by small pteridophytes. The swamp was surrounded by drier land on which grew the cordaites and conifers. Although current ideas differ in detail from Smith's model, his work produced a significant change in our view of the Coal Measure forests; no longer were they seen as a fairly homogeneous flora, but rather as a complex mosaic of plant communities.

THE *GLOSSOPTERIS* FLORA AND CONTINENTAL DRIFT

We now know that all the world's major landmasses were joined together into one super continent called Pangea. Crustal movements split it up and moved the fragments apart to eventually give today's continents. North America was once joined to Europe until the Middle Jurassic. The southern continents were once part of what we call Gondwanaland, which broke up into five fragments that gave South America, South Africa, Australasia, Antarctica and India. All have similar Permian floras with the commonest being a gymnosperm with leaves called *Glossopteris* that was first described and named by Brongniart in 1828 in his *Prodrome d'une Histoire des Végétaux Fossiles* (see Chapter 6). Neymayr was the first to recognize the flora as distinct and called it the *Glossopteris* Flora in 1887 in his first volume of *Erdgeschichte*. Brongniart's specimens came from India and Australia, but *Glossopteris* was not found in South Africa until 1867 and in South America until 1870. All this early work was summarized by Arber in 1905. No specimens from Antarctica were known at that time and it was not until Captain Scott made his fateful second polar expedition in 1911/12 that *Glossopteris* was known from this continent. Specimens collected by the expedition at Mount Buckley at 85° South were among the rock and fossil samples found in Scott's tent and brought back to Britain in the *Terra Nova*. Albert Seward described the plant fossils in 1914, when he also appended his views on the characteristics of the southern flora. Their association with glacial deposits, the small number of species compared to the contemporaneous northern flora, and the dominant part played by *Glossopteris* and the similar *Gangamopteris*. Seward then discussed the distribution of the *Glossopteris* flora and the contemporary glacial features. He dismissed the notion of a shifting axis for the Earth in favour of a considerable change in climate, although the need for a long polar night clearly worried him. Then he returned to the earlier idea of other palaeobotanists that an Antarctic continent was the original area of development of many plant groups and concluded that the discovery of *Glossopteris* added weight to this idea.

Seward, of course, was writing before Wegener had mooted the theory of continental drift in 1915 and long before du Toit (1957) used the distribution of the *Glossopteris* flora to support the argument for continental drift. Much further work has revealed a great deal about the *Glossopteris* plant and the flora, especially important being the discovery of anatomically preserved remains in Australia and Antarctica (Pigg and Trivett, 1994).

Although we have concentrated on the *Glossopteris* flora of Gondwanaland, it should not be thought that there were only two floras in the Upper Palaeozoic. Provincialism had produced four: the Euramerian flora, the Gondwana flora, the Cathaysian flora of Asia and the Far East, and the Angara flora of Siberia and northern Asia, which evolved separately over enormous spans of time. The northern Angara flora is the most different. Studied since the 1840s, it has been shown to consist of simple plants that the Russian Sergei Meyen (1935–1987) described in 1982 as archaic-looking. They probably made very limited or no contribution to the world's Mesozoic floras.

EARLY LAND PLANTS

The first person to give any clear understanding of early land plants was the Canadian, John William Dawson (1820–1899). After studying in Edinburgh, Dawson became Superintendent of Education for Nova Scotia. Through his extensive travels he was able to study geology at first hand, which led in 1855 to his monumental *Acadian Geology*. Dawson then became Principal of the fledgling McGill University. A visit by Sir William Logan, head of the Canadian Geological Survey, encouraged Dawson to study the Devonian plant fossils found along the Gaspé coast. Most of Logan's own extensive collection was lost in a shipwreck, but the few specimens that Dawson saw led him to visit Gaspé to collect more. In 1859 he published his account of *Prototaxites* and *Psilophyton*. Other papers on Devonian plants followed, as did his *Geological History of Plants* in 1888, but they all received little attention. European palaeobotanists interested in the origin of plants just ignored his work, preferring to theorize and base their ideas on plants living today. Their entrenched and introverted ideas held back palaeobotanical studies for decades.

It was not until Robert Kidston and William Lang (1874–?) published their enormously important work on the Devonian Rhynie Chert that palaeobotanists were encouraged to look in Early Palaeozoic rocks for plant fossils (see Chapter 2). Their anatomical descriptions of these simple land plants, published between 1917 and 1921, not only vindicated Dawson's much earlier work, but spurred Lang on to make further studies of Devonian plants. In 1937, Lang established the genus *Cooksonia* that he named after Isabel Cookson with whom he had worked earlier on the Australian *Baragwanathia* flora.

After Kidston and Lang's Rhynie Chert paper appeared, collecting in Gaspé and along the New Brunswick coast was soon undertaken by Loren Petry (1887–1970), Professor of Botany at Cornell University. He was followed in this by other more productive palaeobotanists like Harlan Banks, Chester Arnold and Fran Heuber, who have described a wealth of new and important material from this site – a hundred years or so after Dawson's original paper.

THE AGE OF CYCADS

Research on Mesozoic floras was for many years overshadowed by the overwhelming interest in the Carboniferous. Nevertheless there were people who followed their own interests into the Mesozoic. The earlier people of course were naturalists with much broader interests who combined their study of plant fossils with other studies.

In the early nineteenth century amateur and professional collectors were at work on the Yorkshire coast where they were collecting Jurassic adpressions. The most notable were William Bean and John Williamson (the father of William Williamson referred to earlier). Brongniart described and figured 22 species from Yorkshire, while Lindley and Hutton included 43.

William Buckland (1784–1856) was a priest, Fellow of Corpus Christi College Oxford and later Professor of Mineralogy. He published a number of religious works, especially his Bridgewater Treatise in 1836 in which he set out to prove 'The Power, Wisdom and Goodness of God as manifested in the Creation'. In 1845 he became Dean of Westminster. Meanwhile Buckland spent much time in the Lyme Regis area collecting plant and animal fossils and published many scientific works. Although not many of them were palaeobotanical, he was the first to interpret the quarrymen's silicified 'birds' nests' found at Portland as cycad-like, and in 1828 named them *Cycadeoidea*. In 1840 Buckland even suggested that glaciers had once been in Britain; an idea that was later corroborated by Agassiz. In the same year he was elected President of the Geological Society for the second time.

William Carruthers (1830–1922) was Assistant Keeper and then Keeper of Botany in the British Museum (Natural History). Although his work was mainly on living plants, he did publish a number of accounts on cycads and cycadeoides from the Yorkshire Jurassic, establishing the name *Beania* for a cycadeoid seed. Carruthers named this seed after Mr Bean who had earlier sold a large collection of plant fossils to the museum.

Albert Seward (1863–1941), who became Professor of Botany and then Master of Downing College, was one of the most important palaeobotanists of his time. Seward had studied under Williamson in 1886, but for most of his life he concentrated on Mesozoic fossils, writing monographs on the British Museum (Natural History) collections of Wealden plants between 1894 and 1895 and Jurassic plants between 1900 and 1904 among his range of publications. He also published two great works: the four-volume *Fossil Plants*, between 1898 and 1919, and the popular *Plant Life Through the Ages* in 1933, which gives a readable account of the subject. The rather amazing fact about Seward is that he relied exclusively on museum collections and only collected one British specimen in his life (Harris, 1941). The only exception to this was his collecting trip to Greenland where he was accompanied by Richard Holttum who went on to become a world authority on living ferns. Holttum once said that fossils would not tell him all he wanted to know so after publishing one paper on a fossil fern he left them for others who would find the challenge more stimulating than he did (pers. comm. B.A.T.).

Hugh Hamshaw Thomas (1885–1962) was also at Cambridge for most of his career. He worked briefly with Kidston where he discovered the necessity of collecting and then visited Thore Halle in Sweden. Halle (1884–1964) had visited the Yorkshire coast where he found many of the localities that others had assumed lost or worked out and collected from them. Thomas came back enthused and went to Yorkshire. Once there he found that collecting was not easy and even gave up looking for the famous Gristhope locality, believing it to be lost. Deciding instead to collect seaweed, he tore up a stubborn piece with a piece of attached rock and there were the fossils. An amazing piece of luck because this locality was to provide him with specimens for his most important work. He showed that the seed-bearing *Caytonia*, the pollen-bearing *Antholithius* and the leaves *Sagenopteris* were parts of the same plant and named the group the Caytoniales which he mistakenly believed to be angiosperms (see Chapter 7).

The Yorkshire Jurassic was taken up by Tom Harris (1903–1983) who went to Cambridge to study with Seward. After working on Greenland material and overwintering there while collecting more material, he moved to Reading as Professor of Botany (see Chaloner, 1985). Here, Harris spent most of his research time working on the Jurassic flora of Yorkshire. In his own words, he had 'stolen it from Hamshaw Thomas' (pers. comm. B.A.T.). In 1900, Seward believed that he had described everything of value from the Yorkshire Jurassic and that there was probably nothing worthwhile left to collect. In essence Seward had carried on his researches in the same tradition as Sternberg and Brongniart. Even though he had once prepared a cuticle he decided that they were of no taxonomic value. In contrast, Harris in a long series of papers and his five-volume Yorkshire Jurassic Flora (1961–1979) illustrated how much more can be gained through collecting and using cuticle and spore studies in interpreting the species. Harris was helped in his work by the collections from the Yorkshire coast that had been amassed by Maurice Wonnacott (1902–1990) of the Natural History Museum. After the close of the Second World War Harris visited Yorkshire at least once a year, usually on holiday with his family, but sometimes with other palaeobotanists. He always maintained that there is always more to collect and research and in 1961 he urged people to collect. As he quite rightly suggested, it is usually the collector and not the locality that becomes exhausted because work is continuing on the Yorkshire Jurassic to the present day.

In America Mesozoic plants of a completely different kind were being discovered and they were exciting the public's imagination as never before. In 1849 Lieutenant Simpson of the Unites States Army Corps of Topographical Engineers was accompanying a military expedition into Navajo country in north-eastern Arizona in 1849 when they found 'petrified trees' in what were later shown to be Triassic rocks of the Chinle Formation. Other US Army units later found further large masses of 'petrified trees' in what was subsequently named Lithodendron Wash. The trees, identified as conifers, were described many times, by Simpson and others, and as a result collectors started to come in increasing numbers. Souvenir hunters were soon removing large quantities, as were gem collectors and commercial enterprises. The local population started to complain and, when in the early 1880s a mill was built to crush the petrified wood into abrasives, local indignation boiled over. As a result the Territorial Legislature (Arizona was not yet a state) petitioned Congress to protect the area. In 1900 entry was prohibited and in 1906 some of the area was declared a National Monument. This was the first plant fossil site to receive official state protection. It was subsequently enlarged and on 8 December 1962 was declared the Petrified Forest National Park.

The Black Hills of Dakota yielded further petrified material in 1893, but this time is was fossil cycadeoids that were found. They were studied by George Reber Weiland (1865–1953), who purchased and collected hundreds of specimens that were taken to Yale University. Weiland's work clarified many details, especially the structure of their reproductive organs. His work was carried on by Theodore Delevoryas and William Crepet.

The third important site in the Americas for petrified material is far away from the other two, in Patagonia in southern Argentina. Here there are petrified logs and cones of late Triassic age that once again became collectors' items soon after they were described in the 1920s. Longitudinally cut and polished cones were the favourite and many museums soon had large collections while some are usually found in university teaching collections. Several people, including Weiland, Mary Calder, Rudolph Florin and more recently Ruth Stockey, have researched and published details of them.

FLOWERING PLANTS

For far too long, the origin of the angiosperms was regarded as a complete mystery. 'Ancestral angiosperms' were sought in the Jurassic and Cretaceous but none were found. Fossils of true angiosperms were seen as suddenly appearing in the fossil record, persuading many people to speculate about evolution in the uplands where fossilization was unlikely if not impossible. This belief that plant fossils would supply no effective answer led to many papers being written in the early part of this century on the theoretical aspects of angiosperm evolution. The abandonment of one idea to be followed by another led palaeobotany into disrepute in many people's eyes and plant morphologists sought their answers in comparative morphological studies of living plants. However, in the last thirty, or so, years the balance has been redressed. Now we know much about the evolutionary changes that were occurring within the early angiosperms even if we do not know the precise group from which they originally evolved. When 'mysteries' undergo rigorous scientific investigation they so often turn out to be due to lack of research or understanding.

Angiosperms shed their organs much more than gymnosperms and pteridophytes. Several calculations have been made of the number of the various organs that are shed or abscissed during the life of a large woody angiosperm. A valuable estimate made in 1976 by Norman Hughes of the relative quantities and time available for fossilisation of angiosperm organs quotes: 10^7 pollen grains shed in two months every year, 10^3 flowers shed in one month every year, 10^3 seeds shed in two months every year, 10^5 leaves shed in four months every year, 10^2 twigs shed for two months every year, large woody stems and roots shed only once every ten years.

Obviously all abscissed organs do not become fossilised. Leaves are by far the most common flowering plant fossils found by collectors although pollen can be prepared in enormous numbers from rock samples. However, the first flowering plant fossils to catch the imagination of the early naturalist collectors were not leaves but seeds and fruits that were found pyritised in the Tertiary London Clay. Locals who were collecting pyrite nodules and clay ironstone for sale were scouring these Eocene deposits along the northern shore of the Isle of Sheppey in the Thames Estuary. They also found pyritised seeds, fruits, twigs, fish, crabs and shells and sharks teeth which they started to sell to the interested tourists who came down from London on the river.

In 1757, James Parsons (1705–1770) described and figured some of these seeds and fruits and then, on the basis of their ripeness, deduced that the biblical flood had occurred in the autumn. Over fifty years later, in 1810, Francis Crow described over 700 species that he had amassed over twenty years and concluded that they once belonged to a tropical or high southern latitude vegetation, quite an amazing interpretation considering it was still the beginning of the nineteenth century. But it was in 1841 that James Scott Bowerbank (1799–1877) produced his *A History of the Fossil Fruits & Seeds of the London Clay*. Bowerbank was a well-off amateur palaeontologist who formed the London Clay Club with a number of other enthusiasts. This club gave rise to the Palaeontographical Society, which has survived to the present day and still produces palaeontological monographs.

Little serious work was done on the flora for many years until Mary Reid (1860–1953) and Marjory Chandler (1897–1983) formed their partnership in 1920, which remained unbroken until Reid's death. Together they produced the monumental *London Clay Flora* in 1933 followed in 1961 and 1973 by Chandler's supplements. The work was recently taken up by Margaret Collinson who produced the field guide to the flora in 1983.

Other seed and fruit floras were found and studied on the continent from the beginning of the nineteenth century and both Schlotheim and Sternberg new about them. However, there were no detailed studies, to the level of the work in England, until the middle of this century. Similarly, in the USA studies were spasmodic until as late as the 1950s, when comprehensive studies commenced on the Tertiary Brandon lignite of Vermont.

Leaves of flowering plants are now known from the Cretaceous onwards. For many years it was thought that even the oldest known leaves could be referred to modern families and genera and that they could not provide any evidence about angiosperm evolution. It is only since the early 1970s that more critical studies on leaf architecture and on associated pollen has shown this to be a false premise. Instead, we now know that there was a major Cretaceous adaptive radiation of flowering plants between about 130 and 90 million years ago. Recent studies in North America show that angiosperms appeared in the eastern coastal plains in the latest Barremian or early Aptian and spread rapidly westwards and northwards.

Studies on leaf shape and venation patterns have also since been used to estimate palaeoclimatic limits. Again, the first approach in the 1960s was to use the principle of 'the nearest living relative', that is to make palaeobiological assumptions based upon the environmental parameters of the living taxa. This approach proved very difficult to apply because it relied on very accurate identifications of the fossils and very critical comparisons with living species. A different approach developed in the 1970s uses 'foliar physiogomy' which matches leaf types rather than relying on species identifications. The eight characteristics used are: leaf size distribution, leaf margin type, drip tips, organisation (i.e. simple or compound), venation pattern, venation density, leaf texture and leaf base type.

Fossilised flowers were virtually unknown until fairly recently. It was not really until intense collecting commenced in the 1970s that the recent spate of flowers were found and described in the literature. Now, as we have shown earlier, there are many flowers that have been described either as adpressions or as fusainized three-dimensional remains, with the most extensive collections coming from western Portugal and eastern North America. Another recent parallel development has been the study of pollen grains obtained from the stamens of flowers. This has given important systematic information on pollen types that have been previously recognised only as dispersed grains, some of which were shown in 1996 to be stratigraphically important by Else Marie Friis and Kaj Pedersen.

Phylogenetic analysis (cladistics) has been, of course, used to hypothesise on many aspects of plant and animal evolution but is consistently used to investigate the origin and early diversification of angiosperms. Using this technique Peter Crane, Else Marie Friis and Kaj Pedersen challenged the generally accepted view that the earliest angiosperms were *Magnolia*-like, suggesting instead that they were perhaps herbaceous plants with simple, unisexual flowers lacking differentiation of sepals and petals. There is clearly a great deal left to be considered and future collecting and research is needed, perhaps particularly on Early Cretaceous floras from low palaeolatitudes.

Extinction and disruption at the Cretaceous/Tertiary (K/T) boundary have dominated the research of a number of palaeontologists in recent years. It has been suggested that an asteroid impact was responsible for the extinction of the dinosaurs. Subsequent discoveries of the Iridium Layer and signs of an impact crater in the Yucatan Peninsula of Mexico, dated at about 65 million years ago, have led to this idea becoming the most popular explanation for the mass extinctions of dinosaurs and other groups of large reptiles such as the pterosaurs and pleisiosaurs. The K/T extinction was less profound for plant life. Some recent work has concentrated on the development of vegetation after the event. For example, in 1993 Gary Upchurch and Jack Wolfe used leaf analysis to recognize two major changes: a major increase in precipitation and the development of a zonal distinction between the broad-leaved evergreen vegetation of the southern Rocky Mountains/Gulf Coastal Plain and the broad-leaved deciduous vegetation of the northern Rocky Mountains.

CONSERVATION

Notwithstanding the fact that localities are very rarely worked out, there is the real danger that some may be lost through development or landfill. A few sites such as the Fossil Grove in Glasgow and the Petrified Forest in Arizona have been protected in differing ways for many years, but the vast majority have not been protected. Many localities of palaeobotanical interest, such as working mines or opencasts, can of course never be preserved.

In recent years there has been an increasing desire among geologists to protect the best of the sites. Some countries have initiated schemes for such protection. In

Britain the Geological Conservation Review was initiated in 1977 by the then Nature Conservancy Council to assess, document and publish accounts of the most important geological sites. These GCR selected sites have been included in the lists of Sites of Special Scientific Interest (SSSIs) notified under the Wildlife and Countryside Act 1981. Such listing requires owners and occupiers to notify the relevant government statutory adviser on wildlife and countryside matters if they intend to do anything that may damage the interest of the site. Some sites may be National Nature Reserves, which are fully protected. Volumes on the GCR sites are being published on an interest basis. There will be two palaeobotanical volumes written by us, as well as a number by other authors that will deal with Quaternary sites on a regional basis.

There have been attempts to select geological sites for World Heritage status, but these have failed because there was no international list or database of key earth science sites. The International Union of Geological Societies therefore initiated GEOSITES as a joint IUGS/UNESCO project to compile national and regional inventories of sites from which it should be able to identify those of global significance. Regional working groups have been established throughout the world to develop these inventories. In Britain, for instance, the British Institute for Geological Conservation is undertaking this part of the work.

The situation is rather different in the United States. There are many Federal and State regulations and laws governing the collection of natural history objects. State-owned land is managed by the Department of the Interior, which can control collecting through the issue of permits. Wolberg and Reinhard have recently given a useful summary of the current situation in the United States.

THE FUTURE FOR PALAEOBOTANY

As we have tried to show, palaeobotanists have studied fossil plants for many different reasons. The very first were naturalists in the broad sense who were interested in the variety of life on Earth that they saw around them. Fossil plants were initially studied to show their relationships with living plants. Then in the early nineteenth century the study of fossils became linked to geological investigations, especially the understanding of sedimentary rocks and their sequence in time.

These two different aspects of palaeobotanical study remain with us today, but the differences between them often become blurred through interpretation of evolution, palaeoecology and palaeogeography. Furthermore, the acceptance of continental drift as a reality involving the movements of crustal plates has opened the door to a better understanding of plant distribution and speciation. Our knowledge of palaeoecology and the interactions between animals and plants has increased tremendously in recent years and shows no sign of decreasing in importance for future research work. Similarly, reproductive biology is an ever expanding research field that will continue to give us better insight into whole plant biology.

New techniques introduced in recent years have included biochemical analysis and the use of cladistics as aids to recognising relationships between taxa. Both have given useful and interesting results and their use will undoubtedly increase as new discoveries of plant fossils are made, but where they may lead us in the future is still uncertain.

Palaeobotanists are also using experiments on sedimentation to simulate fossilization in an attempt to understand how the fossil assemblages may relate to original plant communities. From this can come environmental interpretations. Such work is naturally fraught with difficulty because the further back in time we go the less certain we are of our interpretations. Even something seemingly simple like temperature regimes becomes more complicated when latitude and overall global temperature changes are taken into account. Much more integration of different disciplines will be essential if further progress is to be made.

Information can be hard to find, even in this age of fast communication. The literature is ever expanding and the number of journals increasing. There have been attempts to alleviate this problem. The *Fossilium Catalogus* was started in 1913 by Wilhelm Jongmans and edited by him until his death in 1957. It is continuing today and has reached part 99. It is the most comprehensive encyclopaedia of palaeobotany that has been published. There are also bibliographies of American (annual) and European (biennial) palaeobotanical literature.

Good field guides are important assets and many more are becoming available. In Britain the Palaeontological Association has published three on plant fossils: Sheppey Tertiary plants by Margaret Collinson in 1983, Carboniferous plants by Christopher Cleal and Barry Thomas in 1994, and Yorkshire Jurassic plants by Han van Konijnenburg-van Cittert and Helen Morgans in 1999. In North America the most recent are on western North America by Don Tidwell in 1975, West Virginia by Bill Gillespie *et al.* in 1978, Mazon Creek by Raymond Janssen in 1979, the Upper Carboniferous of Nova Scotia by Erwin Zodrow and Keith McCandlish in 1980, and the Pennsylvanian of Illinois by James Jennings in 1990. Such guides are valuable for beginners and amateurs, but it should always be remembered that there is much still to be learned and these guides are not exhaustive. Collecting involves selection and the ultimate rejection of the vast majority of specimens, which are deemed to be duplicates or worthless. Rejects of indeterminable (because they are not in the field guides) status might just be what someone else is looking for. Collecting, within reason, is to be encouraged.

Some regular international meetings help the flow of information. The International Botanical Congress meets every six years and provides the venue for decisions on the taxonomic rules that affect the naming of all plants, including fossils. There is also the Carboniferous Congress (*Congrès Internationale de Stratigraphie et de Géologie du Carbonifère*), mentioned earlier. The meeting of interest to most palaeobotanists is that organized by the International Organization of Palaeobotany (IOP) every four years. The Secretary of the IOP is also currently establishing a Plant Fossil Record database whose long-term objective is to allow immediate international access to descriptive and occurrence details of all plant fossils. This is clearly an enormous task and priority is being given to plants that may have been environmentally sensitive. When completed, the database searches would enable such topics as palaeoecological, palaeoclimatological and extinction rates to be researched with much greater accuracy. Information can be accessed on http://ibs.uel.ac.uk/ibs/

So where should palaeobotany be leading us in the future? In the last twenty to thirty years there have been many outstanding advances in our knowledge of plant fossils. Although the origins of land plants and flowering plants are still not fully understood we have a much clearer picture than was once thought possible. Reconstructions of complete plants have been made in a few instances although this is still impossible for the vast majority of fossils. Anatomical studies, cuticle investigations, pollen and spore recovery from fructifications and biochemical analyses have all increased our overall knowledge of fossil plants. Nevertheless, although many attempts have been made to understand the interrelationships between the various plant groups and the evolutionary pathways that linked them, we still need a lot more information before we are able to do this with any certainty. Obviously in many instances we may never find the crucial fossils upon which we may make interpretations. We must accept that the fossil record is not complete and can never answer all the questions that we have to ask. Logically, very little plant material is ever preserved because the bulk of the dead plant material must be recycled back into the environment. However, new localities are constantly being found, as the remoter areas of the Earth are opened up with easier and safer access. Some will certainly yield exciting discoveries.

Appendix 1

CLASSIFICATION OF VASCULAR PLANTS

The following is a classification of the families of vascular plants that have a reasonably well documented fossil history. There are various classifications currently in use but that used here was published in 1993 as part of *The fossil record 2* (edited M. J. Benton, published Chapman and Hall). The families are mostly shown arranged in their orders, classes and divisions. However, so many orders are currently recognized within the angiosperms that we have instead grouped these families within their respective subclasses, following Cronquist's classification. We have tried to give some idea of the stratigraphical distribution of each family. However, this is only a 'broad-brush' approach, noting the subsystems (i.e. lower ('L.'), middle ('M.') and upper ('U.') parts of systems) in which the families occurred. More detailed information can be obtained from *The fossil record 2*. In some cases, popular names for families are given in square brackets. It must be emphasized that this is not a complete survey of the plant kingdom and that there are many families, especially of angiosperms, that have little or no fossil record and are thus not listed.

Division: Rhyniophyta
   Class: Rhyniopsida
     Order: Rhyniales
       Family: Rhyniaceae (L. Devonian)
   Class: Zosterophyllopsida
     Order: Zosterophyllales
       Family: Zosterophyllaceae (Devonian)
   Class: Horneophytopsida
     Order: Horneophytales
       Family: Horneophytaceae (L. Devonian)
   Class Trimerophytopsida
     Order: Trimerophytales
       Family: Trimerophytaceae (L.–M. Devonian)
Division: Lycophyta
   Class: Lycopsida
     Order: Drepanophycales
       Family: Drepanophycaceae (U. Silurian – U. Devonian)

Classification of vascular plants

  Order: Protolepidodendrales
   Family: Protolepidodendraceae (L. Devonian – L. Permian)
   Family: Eleutherophyllaceae (Carboniferous)
  Order: Lycopodiales
   Family: Lycopodiaceae (M. Devonian – Recent)
  Order Selaginellales
   Family: Selaginellaceae (U. Devonian – Recent)
  Order: Lepidocarpales
   Family: Cyclostigmaceae (U. Devonian)
   Family: Flemingitaceae (Carboniferous)
   Family: Sigillariostrobaceae (Carboniferous)
   Family: Lepidocarpaceae (L. Carboniferous – U. Permian)
   Family: Diaphorodendraceae (U. Carboniferous)
   Family: Spenceritaceae (U. Carboniferous)
   Family: Caudatocarpaceae (U. Carboniferous)
   Family: Pinakodendraceae (U. Carboniferous)
   Family: Sporangiostrobaceae (U. Carboniferous)
   Family: Cyclodendraceae (L. Permian)
   Family: Pleuromeiaceae (Triassic)
  Order: Miadesmiales
   Family: Miadesmiaceae (U. Carboniferous)
  Order: Isoetales
   Family: Isoetaceae (U. Carboniferous – Recent) [Quilworts]
   Family: Chaloneriaceae (U. Carboniferous)
   Family: Takhtajanodoxaceae (L. Triassic)
Division: Sphenophyta
 Class: Equisetopsida
  Order: Pseudoborniales
   Family: Pseudoborniaceae (U. Devonian)
  Order: Bowmanitales
   Family: Bowmanitaceae (U. Devonian – U. Permian) [Sphenophylls]
   Family: Eviostachyaceae (U. Devonian)
   Family: Cheirostrobaceae (L. Carboniferous)
  Order: Equisetales
   Family: Archaeocalamitaceae (L. Carboniferous)
   Family: Calamostachyaceae (L. Carboniferous – U. Permian) [Calamites]
   Family: Tchernoviaceae (U. Carboniferous – U. Permian)
   Family: Gondwanostachyaceae (U. Permian)
   Family: Equisetaceae (U. Permian – Recent)
   Family: Echinostachyaceae (L. Triassic)

Division: Pteridophyta
  Class: Pteropsida
    Order: Cladoxylales
      Family: Cladoxylaceae (M. Devonian – L. Carboniferous)
    Order: Ibykales
      Family: Ibykaceae (M.–U. Devonian)
    Order: Coenopteridales
      Family: Rhacophytaceae (M. Devonian – L. Carboniferous)
      Family: Zygopteridaceae (M. Devonian – L. Carboniferous)
      Family: Stauropteridaceae (M. Devonian – U. Carboniferous)
      Family: Corynepteridaceae (Carboniferous)
      Family: Biscalithecaceae (U. Carboniferous – L. Permian)
    Order: Botryopteridales
      Family: Psalixochlaenaceae (Carboniferous)
      Family: Tedeleaceae (L. Carboniferous – L. Permian)
      Family: Botryopteridaceae (L. Carboniferous – L. Permian)
      Family: Sermeyaceae (U. Carboniferous – U. Permian)
    Order: Urnatopteridales
      Family: Urnatopteridaceae (U. Carboniferous – L. Permian)
    Order: Crossothecales
      Family: Crossothecaceae (U. Carboniferous)
    Order: Marattiales
      Family: Asterothecaceae (L. Carboniferous – U. Permian)
      Family: Marattiaceae (U. Carboniferous – Recent)
    Order: Osmundales
      Family: Osmundaceae (U. Permian – Recent) [Royal Ferns]
    Order: Filicales
      Family: Gleicheniaceae (U. Permian – Recent)
      Family: Cynepteridaceae (U. Triassic)
      Family: Matoniaceae (U. Triassic – Recent)
      Family: Dipteridaceae (U. Triassic – Recent)
      Family: Polypodiaceae (U. Triassic – Recent)
      Family: Dicksoniaceae (U. Triassic – Recent)
      Family: Schizaeaceae (M. Jurassic – Recent)
      Family: Cyatheaceae (L. Cretaceous – Recent)
      Family: Tempskyaceae (Cretaceous)
      Family: Loxsomaceae (L. Cretaceous – Recent)
      Family: Lophosoriaceae (M. Cretaceous – Recent)
    Order: Ophioglossales
      Family: Ophioglossaceae (L. Palaeogene – Recent)
    Order: Marsileales
      Family: Marsileaceae (M. Cretaceous – Recent)

Classification of vascular plants

    Order: Salviniales
      Family: Salviniaceae (U. Cretaceous – Recent)
      Family: Azollaceae (U. Cretaceous – Recent)
Division: Progymnospermophyta
  Class: Progymnospermopsida
    Order: Aneurophytales
      Family: Aneurophytaceae (M.–U. Devonian)
      Family: Protokalonaceae (U. Devonian)
      Family: Protopityaceae (L. Carboniferous)
    Order: Archaeopteridales
      Family: Archaeopteridaceae (M. Devonian – U. Carboniferous)
    Order: Noeggerathiales
      Family: Noeggerathiaceae (U. Carboniferous – L. Permian)
      Family: Tingiostachyaceae (U. Carboniferous – U. Permian)
    Order: Cecropsidales
      Family: Cecropteridaceae (U. Carboniferous)
Division: Gymnospermophyta
  Class: Lagenostomopsida
    Order: Lagenostomales
      Family: Elkinsiaceae (U. Devonian – L. Carboniferous)
      Family: Genomospermaceae (L. Carboniferous)
      Family: Eospermaceae (L. Carboniferous)
      Family: Lagenostomaceae (Carboniferous)
      Family: Physostomaceae (Carboniferous)
  Class: unnamed
    Order: Calamopityales
      Family: Calamopityaceae (L. Carboniferous)
    Order: Callistophytales
      Family: Callistophytaceae (U. Carboniferous – L.Permian)
    Order: Peltaspermales
      Family: Peltaspermaceae (U. Carboniferous – U. Triassic)
      Family: Cardiolepidiaceae (Permian)
      Family: Umkomasiaceae (U. Permian – U. Cretaceous)
    Order: Leptostrobales
      Family: Leptostrobaceae (U. Triassic – M. Cretaceous)
    Order: Glossopteridales
      Family: Arberiaceae (U. Carboniferous – U. Permian) [Glossopterids]
      Family: Caytoniaceae (U. Triassic – U. Cretaceous)
  Order Gigantonomiales
      Family: Emplectopteridaceae (Permian) [Gigantopterids]
  Class: Cycadopsida
    Order: Medullosales
      Family: Trigonocarpaceae (L. Carboniferous – L. Permian)

Family: Potonieaceae (L. Carboniferous – L. Permian)
  Order: Cycadales
    Family: Cycadaceae (L. Permian – Recent) [Cycads]
    Family: Nilsoniaceae (Jurassic)
Class: Gnetopsida
  Order: Bennettitales
    Family: Bennettitaceae (U. Triassic – U. Cretaceous)
    Family: Williamsoniaceae (U. Triassic – U. Cretaceous)
  Order: Pentoxylales
    Family: Pentoxylaceae (L. Jurassic – M. Cretaceous)
  Order: Gnetales
    Family: Gnetaceae (U. Triassic – Recent)
Class: Pinopsida
  Order: Cordaitanthales
    Family: Cordaitanthaceae (L. Carboniferous – L. Permian)
    Family: Rufloriaceae (U. Carboniferous – U. Permian)
    Family: Vojnovskyaceae (Permian)
  Order: Dicranophyllales
    Family: Dicranophyllaceae (U. Carboniferous – U. Permian).
    Family: Trichopityaceae (U. Carboniferous – L. Permian).
  Order: Pinales
    Family: Emporiaceae (U. Carboniferous)
    Family: Buriadiaceae (U. Carboniferous – L. Permian)
    Family: Ferugliocladaceae (L. Permian)
    Family: Utrechtiaceae (Permian)
    Family: Majonicaceae (U. Permian)
    Family: Ullmanniaceae (U. Permian)
    Family: Voltziaceae (L. Triassic – M. Cretaceous)
    Family: Podocarpaceae (L. Triassic – Recent)
    Family: Palissyaceae (U. Triassic – M. Jurassic)
    Family: Araucariaceae (U. Triassic – Recent)
    Family: Pinaceae (U. Triassic – Recent) [Pines]
    Family: Cheirolepidiaceae (U. Triassic – U. Cretaceous)
    Family: Taxaceae (L. Jurassic – Recent) [Yews]
    Family: Pararaucariaceae (M.–U. Jurassic)
    Family: Taxodiaceae (M. Juassic – Recent) [Bald Cypresses]
    Family: Arctopityaceae (U. Jurassic – M. Cretaceous)
    Family: Sciadopityaceae (U. Jurassic – Recent)
    Family: Cephalotaxaceae (M. Jurassic – Recent)
    Family: Cupressaceae (L. Palaeogene – Recent) [Cedars and Junipers]
  Order: Ginkgoales
    Family: Ginkgoaceae (U. Triassic – Recent) [Maidenhair Trees]

Division: Angiospermae (Magnoliophyta)
  Class: Magnoliopsida (Dicotyledons)
    Subclass: Hamamelidae
      Family: Platanaceae (L. Cretaceous – Recent) [Planes]
      Family: Cercidiphyllaceae (U. Cretaceous – Recent)
      Family: Fagaceae (U. Cretaceous – Recent) [Beeches]
      Family: Hamamelidaceae (U. Cretaceous – Recent) [Witch Hazels]
      Family: Ulmaceae (U. Cretaceous – Recent) [Elms]
      Family: Casaurinaceae (L. Palaeogene – Recent)
      Family: Juglandaceae (L. Palaeogene – Recent) [Walnuts]
      Family: Trochodendraceae (L. Palaeogene – Recent)
      Family: Moaceae (M. Palaeogene – Recent) [Mulberries]
      Family: Myricaceae (M. Palaeogene – Recent)
      Family: Urticaceae (M. Palaeogene – Recent) [Nettles]
      Family: Betulaceae (U. Palaeogene – Recent) [Birches]
      Family: Leitneriaceae (L. Neogene – Recent)
    Subclass: Magnoliidae
      Family: Chloranthaceae (L. Cretaceous – Recent)
      Family: Ranunculaceae (L. Cretaceous – Recent) [Buttercups]
      Family: Priscaceae (M. Cretaceous)
      Family: Magnoliaceae (M. Cretaceous – Recent) [Magnolias]
      Family: Lauraceae (M. Cretaceous – Recent) [Laurels]
      Family: Nymphaeaceae (M. Cretaceous – Recent) [Water Lillies]
      Family: Berberidaceae (U. Cretaceous – Recent) [Barberries]
      Family: Menispermaceae (U. Cretaceous – Recent) [Moonseeds]
      Family: Sabiaceae (U. Cretaceous – Recent)
      Family: Ceratophyllaceae (L. Palaeogene – Recent) [Hornworts]
      Family: Annonaceae (M. Palaeogene – Recent) [Custard Apples]
      Family: Aristolochiaceae (M. Palaeogene – Recent) [Birthworts]
      Family: Myristicaceae (M. Palaeogene – Recent) [Nutmegs]
      Family: Saururaceae (M. Palaeogene – Recent)
      Family: Schisandraceae (M. Palaeogene – Recent)
      Family: Illiciaceae (U. Palaeogene – Recent)
      Family: Lardizabalaceae (U. Palaeogene – Recent)
      Family: Sargentodixaceae (U. Palaeogene – Recent)
      Family: Calycanthaceae (L. Neogene – Recent)
      Family: Coriariaceae (L. Neogene – Recent)
      Family: Fumariaceae (L. Neogene – Recent) [Fumitories]
      Family: Piperaceae (L. Neogene – Recent) [Peppers]
    Subclass: Caryophyllidae
      Family: Amaranthaceae (U. Cretaceous – Recent) [Pigweeds]
      Family: Caryophyllaceae (M. Palaeogene – Recent) [Pinks]
      Family: Chenopodiaceae: (L. Neogene – Recent) [Goosefoot]

  Family: Polygonaceae (L. Neogene – Recent) [Knotweeds]
Subclass: Dilleniidae
  Family: Cyrillaceae (U. Cretaceous – Present)
  Family: Theaceae (U. Cretaceous – Recent) [Teas]
  Family: Actinidiaceae (U. Cretaceous – Recent) [Chinese Gooseberries]
  Family: Lecythidaceae (U. Cretaceous – Recent) [Brazil Nuts]
  Family: Pentaphyllaceae (U. Cretaceous – Recent)
  Family: Ericaceae (L. Palaeogene – Recent) [Heaths]
  Family: Salicaceae (L. Palaeogene – Recent) [Willows]
  Family: Bombacaceae (M. Palaeogene – Recent)
  Family: Brassicaceae (M. Palaeogene – Recent) [Mustards]
  Family: Capparaceae (M. Palaeogene – Recent) [Capers]
  Family: Cistaceae (M. Palaeogene – Recent) [Rock-roses]
  Family: Cucurbitaceae (M. Palaeogene – Recent) [Squashes]
  Family: Dipterocarpaceae (M. Palaeogene – Recent)
  Family: Droseraceae (M. Palaeogene – Recent) [Sundews]
  Family: Ebenaceae (M. Palaeogene – Recent) [Ebonies]
  Family: Flacourtiaceae (M. Palaeogene – Recent)
  Family: Malvaceae (M. Palaeogene – Recent) [Mallows]
  Family: Myrsinaceae (M. Palaeogene – Recent)
  Family: Sapotaceae (M. Palaeogene – Recent)
  Family: Sterculiaceae (M. Palaeogene – Recent)
  Family: Styracaceae (M. Palaeogene – Recent)
  Family: Symplocaceae (M. Palaeogene – Recent)
  Family: Elaeocarpaceae (M. Palaeogene – Recent)
  Family: Tiliaceae (M. Palaeogene – Recent) [Basswoods]
  Family: Clusiaceae (U. Palaeogene – Recent) [Garcinias]
  Family: Primulaceae (U. Palaeogene – Recent) [Primroses]
  Family: Violaceae (U. Palaeogene – Recent) [Violets]
  Family: Clethraceae (L. Neogene – Recent) [Pepperbushes]
  Family: Elatinaceae (L. Neogene – Recent) [Water Worts]
  Family: Empetraceae (L. Neogene – Recent) [Crowberries]
  Family: Passifloraceae (L. Neogene – Recent) [Passion Flowers]
  Family: Stachyuraceae (L. Neogene – Recent)
Subclass: Rosidae
  Family: Aceraceae (U. Cretaceous – Recent) [Maples]
  Family: Araliaceae (U. Cretaceous – Recent) [Ginsengs]
  Family: Cornaceae (U. Cretaceous – Recent) [Dogwoods]
  Family: Icacinaceae (U. Cretaceous – Recent)
  Family: Rhamnaceae (U. Cretaceous – Recent) [Buckthorns]
  Family: Rutaceae (U. Cretaceous – Recent) [Rues]
  Family: Sapindaceae (U. Cretaceous – Recent) [Soapberries]
  Family: Fabaceae (L. Palaeogene – Recent) [Peas]

Classification of vascular plants

   Family: Mimosaceae (L. Palaeogene – Recent) [Mimosas]
   Family: Myrtaceae (L. Palaeogene – Recent) [Myrtles]
   Family: Rosaceae (L. Palaeogene – Recent) [Roses]
   Family: Staphyleaceae (L. Palaeogene – Recent) [Bladdernuts]
   Family: Thymelaeaceae (L. Palaeogene – Recent) [Mezerums]
   Family: Vitaceae (L. Palaeogene – Recent) [Grapes]
   Family: Anacardiaceae (M. Palaeogene – Recent) [Cashews]
   Family: Alangiaceae (M. Palaeogene – Recent)
   Family: Apiaceae (M. Palaeogene – Recent) [Parsleys]
   Family: Burseraceae (M. Palaeogene – Recent)
   Family: Buxaceae (M. Palaeogene – Recent) [Boxwoods]
   Family: Celastraceae (M. Palaeogene – Recent) [Bittersweets]
   Family: Euphorbiaceae (M. Palaeogene – Recent) [Spurges]
   Family: Grossulariaceae (M. Palaeogene – Recent) [Gooseberries]
   Family: Hydrangaceae (M. Palaeogene – Recent) [Hydrangeas]
   Family: Linaceae (M. Palaeogene – Recent) [Flaxes]
   Family: Loranthaceae: (M. Palaeogene – Recent)
   Family: Lythraceae (M. Palaeogene – Recent) [Loosestrifes]
   Family: Malpighiaceae (M. Palaeogene – Recent)
   Family: Melastomataceae (M. Palaeogene – Recent)
   Family: Proteaceae (M. Palaeogene – Recent) [Proteas]
   Family: Meliaceae (M. Palaeogene – Recent) [Mahoganies]
   Family: Nyssaceae (M. Palaeogene – Recent)
   Family: Olacaceae (M. Palaeogene – Recent)
   Family: Onagraceae (M. Palaeogene – Recent) [Evening Primroses]
   Family: Rhizophoraceae (M. Palaeogene – Recent) [Mangroves]
   Family: Santalaceae (M. Palaeogene – Recent) [Sandalwoods]
   Family: Simaroubaceae (M. Palaeogene – Recent) [Quasias]
   Family: Aquifoliaceae (U. Palaeogene – Recent) [Hollies]
   Family: Cunoniaceae (U. Palaeogene – Recent)
   Family: Haloragaceae (U. Palaeogene – Recent) [Wild Milfoils]
   Family: Hippuridaceae (U. Palaeogene – Recent) [Mare's Tails]
   Family: Trapaceae (U. Palaeogene – Recent) [Water Chestnuts]
   Family: Combretaceae (L. Neogene – Recent) [Combretums]
   Family: Elaeagnaceae (L. Neogene – Recent) [Oleasters]
   Family: Garyaceae (L. Neogene – Recent) [Silk Tassels]
   Family: Hippocastinaceae (L. Neogene – Recent) [Horse Chestnuts]
   Family: Podostemaceae (L. Neogene – Recent) [River Weeds]
   Family: Viscaceae (L. Neogene – Recent) [Mistletoes]
   Family: Oxalidaceae (U. Neogene – Recent) [Oxalises]
 Subclass: Asteridae
   Family: Oleaceae (L. Palaeogene – Recent) [Olives]
   Family: Apocynaceae (M. Palaeogene – Recent) [Dogbanes]

Family: Asclepiadaceae (M. Palaeogene – Recent) [Milkweeds]
Family: Asteraceae (M. Palaeogene – Recent) [Asters and Dandelions]
Family: Bigoniaceae (M. Palaeogene – Recent) [Bigonias]
Family: Boraginaceae (M. Palaeogene – Recent) [Borages]
Family: Caprifoliaceae (M. Palaeogene – Recent) [Honeysuckles]
Family: Rubiaceae (M. Palaeogene – Recent) [Madders]
Family: Scrophulariaceae (M. Palaeogene – Recent) [Figworts]
Family: Solanaceae (M. Palaeogene – Recent) [Nightshades]
Family: Lamiaceae (U. Palaeogene – Recent) [Mints]
Family: Menythanaceae (U. Palaeogene – Recent) [Bog Beans]
Family: Campanulaceae (L. Neogene – Recent) [Harebells]
Family: Loganiaceae (L. Neogene – Recent) [Loganias]
Family: Pedaliaceae (L. Neogene – Recent) [Pedaliums]
Family: Valerianaceae (L. Neogene – Recent) [Valerians]
Family: Callitrichaceae (U. Neogene – Recent) [Water Starworts]
Family: Plantaginaceae (U. Neogene – Recent) [Plantains]
Class: Liliopsida (Monocotyledons)
Subclass: Arecidae
Family: Araceae (U. Cretaceous – Recent) [Palms]
Family: Arecaceae (U. Cretaceous – Recent) [Philodendrons]
Family: Lemnaceae (U. Palaeogene – Recent) [Duckweeds]
Subclass: Commelinidae
Family: Typhaceae (U. Cretaceous – Recent) [Typhas]
Family: Zingiberaceae (U. Cretaceous – Recent) [Gingers]
Family: Cyperaceae (L. Palaeogene – Recent) [Sedges]
Family: Poaceae (L. Palaeogene – Recent) [Grasses]
Family: Juncaceae (M. Palaeogene – Recent) [Rushes]
Family: Musaceae (M. Palaeogene – Recent) [Bananas]
Family: Sparganiaceae (M. Palaeogene – Recent) [Bur-reeds]
Family: Commelinaceae (L. Neogene – Recent) [Spiderworts]
Family: Xyridaceae (L. Neogene – Recent) [Yellow-eyed Grasses]
Subclass: Alismatidae
Family: Hydrocharitaceae (L. Palaeogene – Recent) [Water Soldiers]
Family: Alismataceae (M. Palaeogene – Recent) [Arrowheads]
Family: Potamogetonaceae (M. Palaeogene – Recent) [Pondweeds]
Family: Butomaceae (U. Palaeogene – Recent) [Flowering Rushes]
Family: Najadaceae (U. Palaeogene – Recent) [Water Nymphs]
Family: Zannichelliaceae (L. Neogene – Recent) [Grass Wracks]
Subclass: Liliidae
Family: Pontederiaceae (M. Palaeogene – Recent) [Pickerel Weeds]
Family: Taccaceae (U. Palaeogene – Recent)

Appendix 2

FURTHER READING

The following is a list of some of the more important books and papers that touch on topics dealt with in this book. It is not a comprehensive bibliography, but should provide the reader with a lead into the literature.

CHAPTER ONE (INCLUDING GENERAL WORKS)

Cleal, C. J. (ed.), 1991. *Plant fossils in geological investigation: the Palaeozoic.* Ellis Horwood, Chichester, 233 pp.

Cleal, C. J. and Thomas, B. A., 1995. *Palaeozoic palaeobotany of Great Britain.* Chapman and Hall, London, xii + 295 pp.

Meyen, S. V., 1987. *Fundamentals of palaeobotany.* Chapman and Hall, London, xxii + 432 pp.

Schopf, J. M., 1975. Modes of fossil preservation. *Review of Palaeobotany and Palynology,* 20, 27–35.

Stace, C. A., 1989. *Plant taxonomy and biosystematics (second edition).* Edward Arnold, London, viii + 264 pp.

Stearn, W. T., 1992. *Botanical Latin (fourth edition).* David and Charles, Newton Abbot, xiv + 546 pp.

Stewart, W. N. and Rothwell, G. W., 1993. *Paleobotany and the evolution of plants (second edition).* Cambridge University Press, Cambridge, xii + 521 pp.

Taylor, T. N. and Taylor, E. L., 1993. *The biology and evolution of fossil plants (second edition).* Prentice Hall, Englewood Cliffs NJ, xxii + 982 pp.

Thomas, B. A., 1981. *The evolution of plants and flowers.* Peter Lowe, London, 116 pp.

Thomas, B. A. and Cleal, C. J., 1993. *The Coal Measures Forests.* National Museum of Wales, Cardiff, 32 pp.

Thomas, B. A. and Cleal, C. J., 1998. *Food of the dinosaurs.* National Museums and Galleries of Wales, Cardiff, 32 pp.

Thomas, B. A. and Spicer, R. A., 1987. *The evolution and palaeobiology of land plants.* Croom Helm, London, x + 309 pp.

White, M. E., 1986. *The greening of Gondwana.* Reed, Frenchs Forest NSW, 256 pp.

CHAPTER TWO

Beck, C. B., 1960. The identity of *Archaeopteris* and *Callixylon*. *Brittonia*, 12, 351–368.

Beck, C. B. and Wight, D. C., 1988. Progymnosperms. *In* Beck, C. B. (ed.), *Origin and evolution of gymnosperms*. Columbia University Press, New York, pp. 1–84.

Chaloner, W. G. and Macdonald, P., 1980. *Plants invade the land*. HMSO, Edinburgh, 18 pp.

Edwards, D., 1994. Towards an understanding of pattern and process in the growth of early vascular plants. *In* Ingram, D. S. and Hudson, A. (eds), Shape and form in plants and fungi. *Linnean Society Symposium Series*, 16, 39–59.

Edwards, D., 1996. New insights into early land ecosystems: a glimpse of a lilliputian world. *Review of Palaeobotany and Palynology*, 90, 159–174.

Edwards, D., 1997. Charting diversity in early land plants: some challenges for the next millenium. *In* Iwatsuki, K. and Raven, P. H. (eds), *Evolution and diversification of land plants*. Springer, Tokyoa, pp. 3–26.

Edwards, D., Davies, K. L. and Axe, L., 1992. A vascular conducting strand in the early land plant *Cooksonia*. *Nature*, 357, 683–685.

Gensel, P. G. and Andrews, H. N., 1984. *Plant life in the Devonian*. Praeger, New York, 381 pp.

Hemsley, A. R., 1990. *Parka decipiens* and land plant spore evolution. *Historical Biology*, 4, 39–50.

Hueber, F. M., 1968. *Psilophyton*: the genus and the concept. *In* Oswald, D. H. (ed.), *Symposium on the Devonian System, Volume 2*. Alberta Society of Petroleum Geologists, Calgary, pp. 815–822.

Hueber, F. M., 1992. Thoughts on the early lycopsids and zosterophylls. *Annals of the Missouri Botanical Gardens*, 79, 474–499.

Kenrick, P., 1994. Alternation of generations in land plants: new phylogenetic and palaeobotanical evidence. *Biological Review*, 69, 293–330.

Kidston, R. and Lang, W. H., 1917–1921. On Old Red Sandstone plants showing structure, from the Rhynie Chert Bed, Aberdeenshire. *Transactions of the Royal Society of Edinburgh*, 51, 761–784 (part 1); 52, 603–627, 643–680, 831–854, 855–902 (parts 2–5).

Niklas, K. J. and Banks, H. P., 1990. A reevaluation of the Zosterophyllophytina with comments on the origin of lycopods. *American Journal of Botany*, 77, 274–283.

Remy, W., Gensel, P. G. and Hass, H., 1993. The gametophyte generation of some Early Devonian land plants. *International Journal of Plant Science*, 154, 35–58.

CHAPTER THREE

Brack-Hanes, S. D. and Thomas, B. A., 1983. A re-examination of *Lepidostrobus* Brongniart. *Botanical Journal of the Linnean Society*, 86, 125–133.

Cleal, C. J., 1993. Pteridophyta. *In* Benton, M. J. (ed.), The fossil record 2. Chapman and Hall, London, pp. 779–794.

DiMichele, W. A. and Skog, J. E. (eds), 1992. The Lycopsida: a symposium. *Annals of the Missouri Botanical Garden*, 79, 447–736.

Gastasldo, R. A., 1986. An explanation for lycopod configuration, 'Fossil Grove' Victoria Park, Glasgow. *Scottish Journal of Geology*, 22, 77–83.

Habgood, K. S., Hemsley, A. R. and Thomas, B. A., 1998. Experimental modelling of the dispersal of *Lepidocarpon* based on experiments using reconstructions. *Review of Palaeobotany and Palynology*, 102, 101–114.

Macgregor, M. and Walton, J., 1955. *The story of fossil grove*. City of Glasgow Public Parks, 32 pp.

Rothwell, G. W., 1984. The apex of *Stigmaria* (Lycopsida), rooting organ of Lepidodendrales. *American Journal of Botany*, 71, 1031–1034.

Thomas, B. A., 1978. Carboniferous Lepidodendraceae and Lepidocarpaceae. *The Botanical Review*, 44, 321–364.

Thomas, B. A., 1997. Upper Carboniferous herbaceous lycopsids. *Review of Palaeobotany and Palynology*, 95, 129–153.

CHAPTER FOUR

Bateman, R. M., 1991. Palaeobiological and phylogenetic implications of anatomically-preserved *Archaeocalamites* from the Dinantian of Oxroad Bay and Loch Humphrey Burn, southern Scotland. *Palaeontographica, Abteilung B*, 223, 1–59.

Batenburg, L. H., 1977. The *Sphenophyllum* species in the Carboniferous flora of Holz (Westphalian D, Saar Basin, Germany). *Review of Palaeobotany and Palynology*, 24, 69–99.

Batenburg, L. H., 1981. Vegetative anatomy and ecology of *Sphenophyllum zwickaviense, S. emarginatum*, and other "compression species" of *Sphenophyllum*. *Review of Palaeobotany and Palynology*, 32, 275–313.

Baxter, R. W., 1963. *Calamocarpon insignis*, a new genus of heterosporous petrified calamiean cone from the American Carboniferous. *American Journal of Botany*, 50, 469–476.

Bureau, E., 1964. *Traité de paléobotanique, 3. Sphenophyta, Noeggerathophyta*. Masson et Cie, Paris, 544 pp.

Cleal, C. J., 1993. Pteridophyta. *In* Benton, M. J. (ed.), The fossil record 2. Chapman and Hall, London, pp. 779–794.

Gastaldo, R. A., 1981. Taxonomic considerations for Carboniferous coalified compression equisetalean strobili. *American Journal of Botany*, 68, 1319–1324.

Good, C. W., 1971. The ontogeny of Carboniferous articulates: calamite leaves and twigs. *Palaeontographica, Abteilung B*, 113, 137–158.

Good, C. W., 1975. Pennsylvanian-age calamitean cones, elater-bearing spores, and associated vegetative organs. *Palaeontographica, Abteilung B*, 153, 28–99.

Harris, T. M., 1931. The fossil flora of Scoresby Sound, east Greenland. Part 1: Cryptogams (exclusive of Lycopodiales). *Meddelelser om Grønland*, 85(2), 3–102.

Mamay, S. H. and Bateman, R. M., 1991. *Archaeocalamites lazarii*, sp. nov.: the range of Archaeocalamitaceae extended from the lowermost Pennsylvanian to the mid-Lower Permian. *American Journal of Botany*, 78, 489–496.

Rayner, R. J., 1992. *Phyllotheca*: the pastures of the Late Permian. *Palaeogeography, Palaeoclimatology, Palaeoecology*, 92, 31–40.

CHAPTER FIVE

Brousmiche, C., 1983. Les fougères sphénoptéridiennes du bassin houiller Sarro-Lorraine. *Publication Société Géologique du Nord*, 10, 1–480.

Camus, J. M., Jermy, A. C. and Thomas, B. A., 1991. *A world of ferns*. Natural History Museum, London.

Camus, J. M., Gibby, M. and Johns, R. J. (eds), 1996. *Pteridology in perspective*. Royal Botanic Gardens, Kew, xx + 700 pp.

Cleal, C. J., 1993. Pteridophyta. *In* Benton, M. J. (ed.), The fossil record 2. Chapman and Hall, London, pp. 779–794.

Collinson, M. E., 1980. A new multiple-floated *Azolla* from the Eocene of Britain, with a brief review of the genus. *Palaeontology*, 23, 213–229.

Dyer, A. F. and Page, C. N. (eds), 1985. Biology of pteridophytes. *Proceedings of the Royal Society of Edinburgh, Section B*, 86, 1–474.

Eggert, D. A. and Delevoryas, T., 1967. Studies on Paleozoic ferns: *Sermeya*, gen. nov. and its bearing on filicalean evolution in the Paleozoic. *Palaeontographica, Abteilung B*, 120, 169–180.

Eggert, D. A. and Taylor, T. N., 1966. Studies of Paleozoic ferns: on the genus *Tedelea* gen. nov. *Palaeontographica, Abteilung B*, 118, 52–73.

Harris, T. M., 1961. *The Yorkshire Jurassic flora, I. Thallophytes – pteridophytes*. British Museum (Natural History), London.

Manchester, S. R. and Zavada, M. S., 1987. *Lygodium* foliage with intact sorophores from the Eocene of Wyoming. *Botanical Gazette*, 148, 392–399.

Millay, M. A., 1997. A review of permineralised Euramerican Carboniferous tree ferns. *Review of Palaeobotany and Palynology*, 95, 191–209.

Miller, C. N., 1971. Evolution of the fern family Osmundaceae based on anatomical studies. *Contributions from the Museum of Paleontology, The University of Michigan*, 23, 105–169.

Read, C. B. and Brown, R. W., 1937. American Cretaceous ferns of the genus *Tempskya*. *United States Geological Survey, Professional Paper*, 186–F, 105–129.

Rothwell, G. W., 1994. Phylogenetic relationships among ferns and gymnosperms; an overview. *Journal of Plant Research*, 107, 411–416.

Rothwell, G. W., 1996. Pteridophytic evolution: an often underappreciated phytological success story. *Review of Palaeobotany and Palynology*, 90, 209–222.

Rothwell, G. W. and Stockey, R. A., 1991. *Onoclea sensibilis* in the Paleocene of North America, a dramatic example of structural and ecological stasis. *Review of Palaeobotany and Palynology*, 709, 113–124.

Surange, K. R., 1966. *Indian fossil pteridophytes*. Council for Scientific and Industrial Research, New Delhi.

Thomas, B. A., 1991 The study of fossil ferns. *In* J. M. Camus (ed.), The history of British pteridology 1891–1991. *The British Pteridological Society Special Publication*, 4. 7–15.

Tidwell, W. D. and Ash, S. R., 1994. A review of selected Triassic to early Cretaceous ferns. *Journal of Plant Research*, 107, 417–442.

CHAPTERS SIX AND SEVEN

Beck, C. B. (ed.), 1988. *Origin and evolution of gymnosperms*. Columbia University Press, New York, xiv + 504 pp.

Bhatnagar, S. P. and Moitra, A., 1996. *Gymnosperms*. New Age International, New Delhi, vi + 467 pp.

Biswas, C. and Johri, B. M., 1997. *The gymnosperms*. Springer, Berlin, and Narosa, New Delhi.

Chandra, S. and Surange, K. R., 1979. Revision of the Indian species of *Glossopteris*. *Birbal Sahni Monograph*, 2, 291 pp.

Cleal, C. J., 1993. Gymnospermophyta. *In* Benton, M. J. (ed.), The fossil record 2. Chapman and Hall, London, pp. 795–808.

Florin, R., 1951. Evolution of cordaites and conifers. *Acta Horti Bergiani*, 15, 285–388.

Florin, R., 1958. On Jurassic taxads and conifers from northeastern Europe and eastern Greenland. *Acta Horti Bergiani*, 17, 259–388.

Gao Zhifeng and Thomas, B. A., 1989. A review of fossil cycad evidence of *Crossozamia* Pomel and its associated leaves from the Lower Permian of Taiyuan, China. *Review of Palaeobotany and Palynology*, 60, 205–223.

Harris, T. M., 1961. The fossil cycads. *Palaeontology*, 4, 313–323.

Harris, T. M., 1964–1979. *The Yorkshire Jurassic flora. Parts II–V.* British Museum (Natural History), London.

Harris, T. M., 1976. The Mesozoic gymnosperms. *Review of Palaeobotany and Palynology*, 21, 119–134.

Kerp, J. H. F., 1988. Aspects of Permian palaeobotany and palynology. X. The west- and central-European species of the genus *Autunia* Krasser emend. Kerp (Peltaspermaceae) and the form-genus *Rhachiphyllum* Kerp (callipterid foliage). *Review of Palaeobotany and Palynology*, 54, 249–360.

Laveine, J.-P., Lemoigne, Y. and Zhang Shanzhen, 1993. General characteristics and paleobiogeography of the Parispermaceae (genera *Paripteris* Gothan and *Linopteris* Presl), pteridosperms from the Carboniferous. *Palaeontographica, Abteilung B*, 230, 81–139.

Long, A. G., 1960. On the structure of *Calymmatotheca kidstoni* Calder (emended) and *Genomosperma latens* gen. et sp. nov. from the Calciferous Sandstone Series of Berwickshire. *Transactions of the Royal Society of Edinburgh*, 64, 29–44.

Mamay, S. H., 1976. Paleozoic origin of cycads. *United States Geological Survey Professional Paper*, 934, 1–48.

Miller, C. N., 1977. Mesozoic conifers. *Botanical Review*, 43, 218–271.

Miller, C. N., 1982. Current status of Paleozoic and Mesozoic conifers. *Review of Palaeobotany and Palynology*, 37, 99–114.

Pigg, K. B. and Trivett, M. L., 1994. Evolution of the glossopterid gymnosperms from Permian Gondwana. *Journal of Plant Research*, 107, 461–477.

Poort, R. J. and Kerp, J. H. F., 1990. Aspects of Permian palaeobotany and palynology XI. On the recognition of true peltasperms in the Upper Permian of western and central Europe and a reclassification of species formerly included in *Peltaspermum*. *Review of Palaeobotany and Palynology*, 63, 197–225.

Rothwell, G. W., 1981. The Callistophytales (Pteridospermopsida). Reproductively sophisticated gymnosperms. *Review of Palaeobotany and Palynology*, 32, 103–121.

Rothwell, G. W., 1982. New interpretation of the earliest conifers. *Review of Palaeobotany and Palynology*, 37, 7–28.

Serbet, R. and Rothwell, G. W., 1992. Characterizing the most primitive seed ferns. I. A reconstruction of *Elkinsia polymorpha*. *International Journal of Plant Science*, 153, 602–621.

Stidd, B. M., 1981. The current status of medullosan seed ferns. *Review of Palaeobotany and Palynology*, 32, 63–101.

Taylor, T. N. and Millay, M. A., 1979. Pollination biology and reproduction in early seed plants. *Review of Palaeobotany and Palynology*, 27, 329–355.

Taylor, T. N. and Millay, M. A., 1981. Morphologic variability of Pennsylvanian lyginopterid seed ferns. *Review of Palaeobotany and Palynology*, 32, 27–62.

Thomas, H. A. and Bancroft, N., 1913. On the cuticles of some recent and fossil cycadean fronds. *Transactions of the Linnean Society (Botany)*, 8, 155–204.

Tralau, H., 1968. Evolutionary trends in the genus *Ginkgo*. *Lethaia*, 1, 63–101.

Wang Ziqiang, 1997. Permian *Supaia* fronds and an associated *Autunia* fructification from Shanxi, China. *Palaeontology*, 40, 245–277.

Watson, J. and Sincock, C. A., 1992. *Bennettitales of the English Wealden*. Palaeontographical Society, London, 228 pp.

CHAPTER EIGHT

Beck, C. B. (ed.), 1976. *Origin and early evolution of angiosperms*. Columbia University Press, 341 pp.

Collinson, M. E., 1983. *Fossil plants of the London Clay*. Palaeontological Association Field Guide to Fossils, No. 1. 121 pp.

Collinson, M. E., Boulter, M. C. and Holmes, P. L., 1993. Magnoliophyta ('Angiospermophyta'). *In* Benton, M. J. (ed.), The fossil record 2. Chapman and Hall, London, pp. 809–841.

Crane, P. R., 1982. Betulaceous leaves and fruits from the British Upper Palaeocene. *Botanical Journal of the Linnean Society*, 83, 103–36.

Crane, P. R., 1989. Paleobotanical evidence on the early radiation of nonmagnoliid dicotelydons. *Plant Systematics and Evolution*, 162, 165–191.

Crane, P. R. and Blackmore, S., 1989. *Evolution, systematics, and fossil history of the Hamamelidae*. Systematics Association, Special Volume 40, xxiv + 661 pp (2 vols).

Crane, P. R. and Dilcher, D. L., 1985. *Lesqueria*: an early angiosperm fruiting axis from the mid-Cretaceous. *Annals of the Missouri Botanical Garden*, 71, 384–402.

Friis, E. M., Chaloner, W. G. and Crane, P. R. (eds), 1987. *The origins of angiosperms and their biological consequences*. Cambridge University Press, x + 358.

Friis, E. M., Crane, P. R. and Pedersen, K. R., 1997. Fossil history of magnoliid angiosperms. *In* Iwatsuki, K. and Raven, P. H. (eds), *Evolution and diversification of land plants*. Springer, Tokyoa, pp. 121–156.

Friis, E. M. and Endress, P. K., 1990. Origin and evolution of angiosperm flowers. *Advances in Botanical Research*, 17, 99–162.

Friis, E. M. and Skarby, A., 1982. *Scandianthus* gen. nov., angiosperm flowers of saxifragalean affinity from the Upper Cretaceous of southern Sweden. *Annals of Botany*, 50, 569–583.

Herendeen, P. S. and Dilcher, D. L. (eds), 1992. *Advances in legume systematics part 4. The fossil record.* The Royal Botanic gardens, Kew.

Hughes, N. F., 1994. *The enigma of angiosperm origins.* Cambridge University Press, xiii + 303 pp.

Labandiera, C. C., Dilcher, D. L., Davies, D. R. and Wagner, D. L., 1994. Ninety-seven million years of angiosperm-insect association: paleobiological insights into the meaning of coevolution. *Proceedings of the National Academy of Sciences, USA*, 91, 122278–122282.

Manchester, S. R., 1987. The fossil history of the Juglandaceae. *Monographs in Systematic Botany, Missouri Botanic Garden*, 21, 1–137.

Manchester, S. R., 1992. Flowers, fruits, and pollen of *Florissantia*, an extinct malvalean genus from the Eocene and Oligocene of North America. *American Journal of Botany*, 79, 996–1008.

Taylor, D. W. and Hickey, L. J. (eds), 1996. *Flowering plant origin, evolution & phylogeny.* Chapman and Hall, New York.

Tidwell, W. D. and Nambudiri, E. M. V., 1989. *Tomlinsonia thomassonii*, gen. et sp. nov., a permineralized grass from the Upper Miocene Ricardo Formation, California. *Review of Palaeobotany and Palynology*, 60, 165–177.

CHAPTER NINE

Behrensmeyer, A. K., Damuth, J. D., DiMichele, W. A., Potts, R., Sues, H.-D. and Wing, S. L. (eds), 1992. *Terrestrial ecosystems through time.* University of Chicago Press, Chicago.

Boulter, M. C. and Fisher, H. C. (eds), *Cenozoic plants and climates of the arctic.* Springer Verlag, viii + 401 pp.

Cleal, C. J. (ed.), 1991. *Plant fossils in geological investigation: the Palaeozoic.* Ellis Horwood, Chichester, 233 pp.

Collinson, M. E., Fowler, K. and Boulter, M. C., 1981. Floristic changes indicate a cooling climate in the Eocene of southern England. *Nature*, 291, 315–317.

Dobruskina, I. A., 1994. Triassic floras of Eurasia. *Österrieiche Akademie der Wissenschaften, Schriftenreihe der Erdwissenschaften Kommissionen*, 10, 422 pp.

Gensel, P. G. and Andrews, H. N., 1984. *Plant life in the Devonian.* Praeger, New York, 381 pp.

Meyen, S. V., 1987. *Fundamentals of palaeobotany.* Chapman and Hall, London, xxii + 432 pp.

Vakhrameev, V. A., 1991. *Jurassic and Cretaceous floras and climates of the Earth.* Cambridge University Press, Cambridge, xix + 318 pp.

Vakhrameev, V. A., Dobruskina, I. A., Meyen, S. V. and Zaklinskaja, E. D., 1978. *Paläozoische und mesozoische Floren Eurasiens und die Phytogeographie dieser Zeit.* Gustav Fischer, Jena, 300 pp.

Wolfe, J. A., 1997. Relations of environmental change to angiosperm evolution during the Late Cretaceous and Tertiary. *In* Iwatsuki, K. and Raven, P. H. (eds), *Evolution and diversification of land plants.* Springer, Tokyoa, pp. 269–290.

CHAPTER TEN

Andrews, H. N., 1980. *The fossil hunters. In search of Ancient Plants*. Cornell University Press, Ithica, 421 pp.

Arber, E. A. N., 1905. *Catalogue of the fossil plants of the Glossopteris flora in the Department of Geology, British Museum (Natural History)*, London. Lxxxiv + 255 pp.

Ash, S. R., 1972. The search for plant fossils in the Chinle Formation. *In* Breed, C. R. and Breed W.J. (eds), *Investigations in the Triassic Chinle Formation*. Museum of North Arizona Bulletin, 47, 45–58.

Campbell, S and Bowen, D. Q., 1989. *Quaternary of Wales*. Nature Conservancy Council., Peterborough. 237 pp.

Cleal, C. J. and Thomas, B. A., 1994. *Plant fossils of the British Coal Measures*. Palaeontological Association Field Guide to Fossils, No. 6. 222 pp.

Cleal, C. J. and Thomas, B. A., 1995. *Palaeozoic palaeobotany of Great Britain*. Chapman and Hall, London. 295 pp.

Cleal, C. J. and Thomas, B. A. (in press). *Mesozoic and Tertiary palaeobotany of Great Britain*. Chapman and Hall, London.

Collinson, M. E., 1983. *Fossil plants of the London Clay*. Palaeontological Association Field Guide to Fossils, No. 1. 121 pp.

Crookall, R., 1955–1976. Fossil plants of the Carboniferous rocks of Great Britain [Second Section]. *Memoirs of the Geological Survey of Great Britain, Palaeontology*, 4(1–7), 1004 pp.

Du Toit, A., 1957. *Our wandering continents*. Oliver Boyd, Edinburgh.

Edwards, D., 1990. Robert Kidston: the most professional palaeobotanist. A tribute on the 60th anniversary of his death. *Forth Naturalist and Historian*, 8, 65–93.

Gillespie, W. H., Clendening, J. A. and Pfefferkorn, H. W., 1978. *Plant fossils of West Virginia*. West Virginia Geological and Economic Survey, Morgantown, West Virginia, 172 pp.

Harris, T. M., 1941. Sir Albert Seward (1863–1941). *Royal Society of London Obituary Notices, Fellows*, 3, 213–220.

Harris, T. M., 1963. Hugh Hamshaw Thomas (1885–1963). *Royal Society of London Obituary Notices, Fellows*, 3, 213–220.

Janssen, R. E., 1979. *Leaves and stems from fossil forests* (4th revised edition). Illinois State Museum, Springfield, Illinois, 190 pp.

Jennings, J. J., 1990. *Guide to the Pennsylvanian fossil plants of Illinois*. Illinois State Geological Survey, Champaign, Illinois, 75 pp.

Kidston, R., 1923–1925. Fossil plants of the Carboniferous rocks of Great Britain [First Section]. *Memoirs of the Geological Survey of Great Britain, Palaeontology*, 2(1–6), 681 pp.

Konijnenburg-van Cittert, J. van and Morgans, H. 1999. *A guide to the Yorkshire Jurassic flora*. Palaeontological Association Field Guide to Fossils.

Kosanke, R. M., 1950. Pennsylvanian spores of Illinois and their use in correlation. *Illinois State Geological Survey Bulletin*, 74, 128 pp.

Long, A. G., 1996. *Hitherto*. The Pentland Press, Edinburgh, 278 pp.

Meyen, S. V., 1982. The Carboniferous and Permian plants of Angaraland (A synthesis). *Biological Memoirs*, 7, 1–110.

Pigg, K. B. and Trivett, M. L., 1994. Evolution of the Glossopterid gymnosperms from Permian Gondwana. *Journal of Plant Research*, 107, 461–477.

Schopf, J. M., Wilson, L. R. and Benthall, R., 1944. An annotated synopsis of Paleozoic fossil spores and the definition of generic groups. *Illinois State Geological Survey, Report of Investigations*, 91, 73 pp.

Seward, A. C., 1914. *British Antarctic ("Terra Nova") Expedition, 1910. Natural History Report. Geology vol. 1. No. 1, Antarctic Fossil Plants*. British Museum (Natural History), London, 49 pp.

Smith, A. H. V., 1964. The palaeoecology of Carboniferous peats based on the miospores and petrography of bituminous coals. *Proceedings of the Yorkshire Geological Society*, 33, 423–474.

Smith, A. H. V. and Butterworth, M., 1967. Miospores in the Coal Measure seams of the Carboniferous of Great Britain. *Special Papers in Palaeontology*, 1, 1–324.

Thomas, B. A., 1986. *In search of fossil plants: the life and work of David Davies (Gilfach Goch)*. National Museum of Wales, 54 pp.

Tidwell, W. D., 1975. *Common fossil plants of western North America*. Brigham Young University Press, Provo, Utah, 197 pp.

Williamson, W. C., 1896. *Reminiscences of a Yorkshire naturalist*. George Redway, London 228 pp. (Reprinted with additions by J. Watson and B. A. Thomas, 1985.)

Wolberg, D. and Reinhard, P., 1997. *Collecting the natural world: legal requirements and personal liability for collecting plants, animals, rocks, minerals and fossils*. Geoscience Press, viii + 330 pp.

Zodrow, E. L. and McCandish, K., 1980. *Upper Carboniferous fossil flora of Nova Scotia*. Nova Scotia Museum, Halifax, Nova Scotia, 275 pp.

PLATE EXPLANATIONS

Plate 1. A transverse section through two stems of the Early Devonian plant *Rhynia gwynnevaughanii* Kidston and Lang (x 40). The petrifactions from the Rhynie Chert provide exquisite detail of the anatomy of this plant, whose stems can be seen to consist of a very slender central vascular strand and a thick cortex. Also clearly visible are irregular protuberances from the surface of the stem, which is a characteristic feature of this plant. This was for many years regarded as the archetypal primitive land plant and is still by far the best known of the rhyniophyte plants. The picture was taken from a microscope slide in the collections of the National Museums and Galleries of Wales, Cardiff (No. 21.14G.13). Photo by B. A. Thomas. Printed by the Photography Department, NMGW.

Plate 2. A longitudinal section through the gametophyte (sexual generation) of a plant from the Lower Devonian Rhynie Chert (x 15). In living vascular plants, the gametophyte is much smaller than the sporophyte (vegetative generation), but in the very early land plants both generations are comparable in size. This type of gametophyte, known as *Lyonophyton rhyniensis* Remy and Remy, is thought to be the male sexual equivalent of *Aglaophyton major* (Kidston and Lang) D. S. Edwards. It consists of a slender axis terminated by a cup-shaped structure, the latter being covered with anteridia on its inner surface. The anteridia, one of which can be seen as a black structure in section on the left-hand side of the cup, have been found with beautifully preserved coiled sperm. The picture was taken from a microscope slide in the collections of the Geologisch-Paläontologisches Institut und Museum, Westfälische Wilhelms-Universität, Münster, Germany. The photo was provided by H. Kerp, Münster.

Plate 3. Fragments of *Cooksonia*, showing the variation in the shape of the terminal sporangia used to classify them into species. This tiny plant was probably the very earliest land-vascular plant. Top left is *C. pertoni* Lang from its type locality at Perton Lane, Herefordshire (x 10). It clearly shows the characteristic flattened sporangia (Specimen No. 77.6G.114). Bottom left is *C. pertoni* Lang from the lower Ludlow Series of Cwm Craig Ddu near Builth Wells, Wales (x 6). Until the discovery of specimens in the Wenlock Series of Ireland, these were the oldest known specimens of this genus known from anywhere in the world (Specimen No. 79.17G.3). Top right is *C. hemispherica* Lang which has more rounded sporangia borne on a gradually widening stem (x 35; Specimen No. 77.6G.27a). Bottom right is *C. cambrensis* Edwards, with rounded sporangia on stems that do not widen (x 15; Specimen No. 77.6G.21). The bottom two specimens come from the Downton Series of Freshwater East, Pembrokeshire. Specimens from the National Museums and Galleries of Wales, Cardiff. Photos by the Photography Department, NMGW.

Plate 4. *Steganotheca striata* Edwards from the upper Ludlow Series of Capel Horeb, near Llandovery, Wales (x 2.7). This plant looked rather like *Cooksonia*, except that it was a little larger and had more elongate sporangia. However, vascular tissue has not been found

Explanations of plates

in the stems, and so it is referred to as a rhyniophytoid rather than a rhyniophyte. Specimen from the National Museums and Galleries of Wales, Cardiff (No. 69.64G.32a). Photo by the Photography Department, NMGW.

Plate 5.   At the time that vascular plants were first evolving, a number of non-vascular plants were also adapting to life on land. This figure shows two examples of a thalloid plant, *Parka decipiens* Fleming (both x 5), which was probably a primitive relative of the bryophytes. They originate from the Lower Devonian of Balgavies Quarry, Forfar, Scotland. Specimens from the Natural History Museum, London (Nos OR42665 and V.57875) Photo A. Hemsley, University of Wales, Cardiff.

Plate 6.   A group of fertile spikes of *Zosterophyllum llanoveranum* Croft and Lang, showing the sporangia borne laterally on the otherwise naked axes (x 2.5). Plants of this genus coexisted with the early rhyniophytes such as *Cooksonia* but were somewhat larger. They gave rise to the club mosses (lycopsids) that became a major component of land vegetation during the Carboniferous. This specimen came from the Lower Devonian of Llanover Quarry, near Abergavenny, Wales, and is now stored in the Natural History Museum, London (No. V.26516a). Photo by the Photographic Studio, Natural History Museum, London.

Plate 7.   Close up of a fertile spike of *Zosterophyllum myretonianum* Penhallow, showing the bivalved sporangia attached laterally to the axis (x 4). Specimen originated from the Lower Devonian of Clocksbriggs Quarry, Forfar, Scotland, and is now stored in the Natural History Museum, London (No. V.58047). Photo by the Photographic Studio, Natural History Museum, London.

Plate 8   *Zosterophyllum* had a tufted mode of growth, with a mass of stems lying along the ground, and giving rise to vertical stems that bore the sporangia. This specimen of *Z. myretonianum* Penhallow from Balgavies Quarry near Forfar, Scotland, shows the basal mass of stems that lay along the ground giving rise to fertile uprights (x 1). The specimen is now stored in the Natural History Museum, London (No. V.58047). Photo by the Photographic Studio, Natural History Museum, London.

Plate 9.   This is a substantial part of a *Gosslingia breconensis* Heard plant, and shows the mixture of lateral and dichotomous branches (x 1.5). Parts of the stems were petrified by pyrite, which preserved some of their internal anatomy, confirming that this plant was a zosterophyll. The specimen originated from the Lower Devonian (Pragian) of Brecon Beacons Quarry, South Wales. Specimen from the National Museums and Galleries of Wales, Cardiff (No. 69.64.G1). Photo by the Photography Department, NMGW.

Plate 10.   Close up of a fertile branch of *Gosslingia breconensis* Heard, showing the sporangia attached laterally along the length of the stem (x 2.4). This contrasts with *Zosterophyllum*, in which the sporangia are clustered into fertile 'spikes' at the end of the stems. The specimen originated from the Lower Devonian (Pragian) of Brecon Beacons Quarry, South Wales. Specimen from the National Museums and Galleries of Wales, Cardiff (No. 69.64.G2). Photo by the Photography Department, NMGW.

Plate 11.   A fertile shoot of *Psilophyton crenulatum* Doran from the Lower Devonian (Emsian) of New Brunswick, Canada (x 4). This compressed shoot was extracted by dissolving away the rock matrix with hydrofluoric acid, to show the three-dimensional configuration of the sporangial clusters, and of the small spines attached to the stem. The clusters of sporangia borne on recurved branches are typical of the trimerophytes, the group of plants that hold an intermediate evolutionary position between the primitive rhyniophytes, and the cladoxylaleans and progymnosperms. Specimen from the University of Alberta Palaeobotanical Collections, Edmonton (No. S7109). Photo J. B. Doran, supplied by P. G. Gensel, University of North Carolina.

Plate 12. A fertile shoot of *Protopteridium thomsonii* (Dawson) Kräusel and Weyland, from the Middle Devonian Sandwick Fish Bed, Bay of Skaill, Orkney (x 4). This is the oldest known progymnosperm, the group of plants that gave rise in the Late Devonian to the seed plants (gymnosperms). These terminal clusters of sporangia somewhat resemble the reproductive organs of the trimerophytes, from which they probably evolved, but were borne on stems or trunks with typically gymnosperm secondary wood (when first discovered in the nineteenth century, it was described as an early example of conifer wood!). The specimen is now stored in the Natural History Museum, London (No. V.9425). Photo by the Photographic Studio, Natural History Museum, London.

Plate 13. *Svalbardia polymorpha* Høeg showing both the deeply segmented leaves (top left) and the fertile branches with numerous clusters of elongate sporangia (bottom right) (x 0.9). This and some other specimens collected during the 1928 Vogt Expedition to Spitsbergen are the most complete known examples of *Svalbardia* ever discovered, and show how this genus had *Archaeopteris*-like sporangial clusters, but more primitive, *Protopteridium*-like foliage. The specimen came from the uppermost part of the Middle Devonian, Planteryggen, Mimerdalen, north-western Spitsbergen. It is now stored in the Paleontologisk Museum, Oslo, Norway (No. PA 335). Photo by Franz-Josef Lindemann, Paleontologisk Museum, Oslo.

Plate 14. This fertile shoot of the progymnosperm *Archaeopteris roemeriana* (Göppert) Lesquereux from the Upper Devonian (Famennian) of Durnal, Belgium (x 1.25). It clearly shows shoots bearing either helically arranged leaves or sporangial clusters. Such a shoot would probably have been borne on a large tree with substantial secondary wood. These plants were probably the ancestors of gymnosperms. Specimen from the Royal Institute of Natural Science of Belgium, Brussels (No. b 2421 a). Photo P. Kenrick, Natural History Museum, London.

Plate 15. When secondary wood such as this was first discovered in the nineteenth century, it was thought to be evidence of very early conifers. However, in the late 1950s it was discovered that such wood in fact belonged to a quite different a now extinct group of plants known as the progymnosperms, which combined gymnosperm-like stems and pteridophytic reproduction. This particular transverse section (x 50) is through a branch of *Callixylon newberryi* (Dawson) Elkins from the Upper Devonian New Albany Shale of southern Indiana, USA and clearly shows growth rings. Similar specimens from here have reached a diameter of 1 m and so clearly belonged to a large tree. The photograph is from a microscope slide in the W. S. Lacey Collection, National Museums and Galleries of Wales, Cardiff (No. 87.77G.589). Photo by B. A. Thomas. Printed by the Photography Department, NMGW.

Plate 16. *Baragwanathia oblongifolia* Lang and Cookson from the Upper Plant Assemblage (Lower Devonian) of Victoria, Australia (x 1.7). This is the oldest known lycophyte species. It also occurs in the stratigraphically lower Lower Plant assemblage of Victoria, which has been thought to be middle Silurian in age, making it one of the oldest land plants from anywhere in the world. However, the age of this lower horizon has been queried and may be approximately the same as this specimen. The specimen is in the Smithsonian Institution, Washington DC (No. USNM 446315). Photo by F. Hueber, supplied by W. A. DiMichele. Printed by R. Green (Smithsonian Institution).

Plate 17.   Transverse sections through several stems of the early lycophyte *Asteroxylon mackei* Kidston and Lang, from the Lower Devonian Rhynie Chert (x 4.5). The fossil clearly shows the star-shaped cross-section of the vascular strands. In one stem, two such strands can be seen, as the section is just below a fork in the stem. In several cases, slender leaves can be seen surrounding the stems, represented by small, oval sections. Near the top of the slide is a slender stem of the same plant, preserved in longitudinal section, showing these leaves in attachment to the stem. The specimen is one of those originally described by Kidston and Lang, and is now in the Hunterian Museum, Glasgow (No. 2579). Photo by the Photography Department, NMGW.

Plate 18.   These are sections through permineralised lycophyte stems showing the comparatively narrow vascular strand surrounded by thick cortical tissues. The upper is *Paralycopodites brevifolius* (Williamson) DiMichele from the Pettycur Limestone (Albian), Pettycur, Scotland (x 4). The lower is *Lepidophloios wuenschianus* (Williamson) Walton from Vissean ash deposits at Laggan on the north-east coast of the Isle of Arran, Scotland (x 2). Both specimens are from the Natural History Museum, London (Nos WC.223 & WC.456a). Photo by the Photographic Studio, Natural History Museum, London.

Plate 19.   The 'paper coal' from the Lower Carboniferous of the Moscow Basin is a mass of naturally macerated lycophyte cuticles. This specimen, called *Eskdalia oliverii* (Auerbach and Trautschold) Thomas, shows a stem cuticle, of simple cells, with perforations marking the former attachment sites of leaves (x 80). Ligule pit cuticles hang from the upper angle of the leaf perforations. Specimen from the National Museums and Galleries of Wales, Cardiff (No. 98.24.G1). Photo B. A. Thomas, printed by the Photography Department NMGW.

Plate 20.   The Upper Devonian arborescent lycophyte *Cyclostigma kiltorkense* Haughton from Kiltorkan, Kilkenny, Ireland is preserved by chlorite mineral replacement. Upper left, a detached truncated cone segment exposed by the plane of cleavage passing over the ends of the sporangia (x 1). Specimen from the Geological Survey, Dublin (No. 3075). Upper right, the same specimen photographed under xylene. A megaspore is visible to the right of the cone (x 1.66). Lower right, a sporangium full of megaspores (x 12). Specimen from the Natural History Museum, London (No. V5934). Lower left, the plant, like other arborescent lycophytes, shed its leaves so that the larger stems show leaf scars such as this on their outer surfaces (x 30). There is a vascular print together with two lateral aerating canals, parichnos, in the upper half of the scar. Specimen from the Geological Survey, Dublin. All photos W. G. Chaloner.

Plate 21.   The leaf cushions of this Upper Carboniferous lycophyte, *Lepidodendron aculeatum* Sternberg, from Radnice, Bohemia, Czech Republic, are broad-rhomboidal with inflexed upper and lower angles (x 2). Keels extend to the upper and lower angles from the central leaf scar and lateral lines curve downwards from its corners. The scar has a vascular print and two parichnos. A ligule scar is adjacent to its upper and two infrafoliar parichnos are just below the leaf scars in the lower part of the cushion. The diagonal lines are cracks in the compression. Specimen from the Prague Museum (No. CGH 658 – the type specimen for *Lepidodendron obovatum* Sternberg which is a synonym of *L. aculeatum*). Photo B. A. Thomas, printed by the Photography Department, NMGW

Plate 22. A dichotomised shoot of *Lepidodendron mannabachense* Presl with leaves still attached to the apical portion of one branch. The specimen comes from the Lower Westphalian Žacleř Formation, Intra-Sudetic Coalfield, Poland, and is now in the collections of the Museum für Naturkunde, Berlin. Photo B. A. Thomas. Printed by the Photography Department, NMGW.

Plate 23. This portion of lepidodendroid shoot (x 3) from the British Coal Measures bears knobs in spirals that mark the position of former branches. Such specimens are often given the name of *Halonia*. Specimen from the National Museums and Galleries of Wales, Cardiff (No. 68.163G1). Photo by the Photography Department, NMGW.

Plate 24. A cut coal ball from the Upper Carboniferous of Lancashire, England (x 1.3). Horizontally elongated leaf cushions of the arborescent lycophyte *Lepidophloios* are visible on the surface (turn the page to have the cut surface on the top). The outer tissues of this *Lepidophloios laricinus* Sternberg stem are preserved in the coal ball. Specimen from the National Museums and Galleries of Wales, Cardiff (No. 98.24.G2). Photo by the Photography Department, NMGW.

Plate 25. A section of a similar coal ball from the Upper Foot Mine, Shore, Littleborough, Greater Manchester. It shows a section through a *Lepidophloios* stem with its small central vascular cylinder, thickened outer cortex and swollen leaf cushions (x 3). The inner cortex decayed and was invaded by rootlets of *Stigmaria* before the stem was petrified. The specimen is in the collections of the National Museum of Wales, Cardiff (No. 21.77.15). Photo by the Photography Department NMGW.

Plate 26. Victoria Park in Glasgow has the best known examples of *in situ* stumps of arborescent lycopsids in the Lower Carboniferous, some 325 million years ago. The rooting structures are about 0.4 m tall and clearly of the *Stigmaria* type although much of their narrower dichotomising parts have not been preserved. The stumps provide an unique insight into the forests which were starting to dominate the tropical region at that time and allow us to estimate that the tree density in the forests at that time to be about 4500 per km^2. Such stands of lycophyte stumps are relatively common in the Upper Carboniferous. Photo B. A. Thomas.

Plate 27. The upper picture shows the impressive *Stigmaria* that was excavated from an Upper Carboniferous quarry near Bradford, Yorkshire and taken to the Manchester Museum in 1886 by Professor W. C. Williamson. It was the first such specimen of *Stigmaria* to be found and was illustrated by Williamson in his monographic publication explaining the structure of these rooting bases. The lower specimen of *Stigmaria ficoides* Brongniart, from above the Pentre Coal, Gilfach Goch, South Wales, shows a small section of a narrow axis to the left and the true roots attached at right angles to the right. The attachment points of the roots can be seen as circular prints all over the stigmarian axis. This specimen is from the National Museum and Galleries of Wales, Cardiff (No. NMW

22.113G97). Upper photo provided by Joan Watson and printed by the Photography Department, NMGW; lower photo by the Photography Department, NMGW.

Plate 28.   Cross section of the woody cylinder of *Stigmaria ficoides* Brongniart from a coal ball out of the Halifax Hard Coal (Langsettian) of Yorkshire. The rows of secondary wood tracheids radiate out from the primary xylem, which was original circular and surrounding a central pith. Slight compression prior to petrifaction has distorted the symmetry. Specimen in the collections of the National Museums and Galleries of Wales, Cardiff (No. 60.538.9). Photo by the Photography Department NMGW.

Plate 29.   Top: Model of apical portion of *Stigmaria* showing the roots emerging at right angles. Compare with the fossils shown on Pl. 27. Bottom: Model of outer tissues of a *Lepidodendron* trunk showing, various stages of leaf abscission and decortication. From left to right: leafy shoot, leaf cushions after foliar abscission, loss of leaf cushions, loss of outer cortex. Both models are the work of Annette Townsend and are now in the collections of the National Museums and Galleries of Wales, Cardiff (V97.26.9, V97.26.4). Photos by the Photography Department, NMGW.

Plate 30.   A reconstructed view through an arborescent lycophyte forest in a Carboniferous, Coal Measures, floodplain. Various stages in growth are visible from the leafy unbranched pole-like younger plants through the branching apical phase to the fully-grown fertile plant in the foreground with its terminal cones. Painting by Annette Townsend in the collections of the National Museums and Galleries, Cardiff. Photo by the Photography Department, NMGW.

Plate 31.   This lycophyte stem, *Sigillaria nortense* Crookall, from the upper Westphalian D of Kilmersdon Colliery Tip, near Radstock, England, has its spiral leaf scars level with the stem surface and secondary arranged into vertical rows on ribs (x 1.5). The vascular scar and two curved parichnos prints are clearly visible on all the leaf scars. The specimen is an impression so the grooves between the ribs appear as ridges. *Sigillaria*, unlike *Lepidodendron*, rarely branches. The specimen is from the National Museums and Galleries of Wales, Cardiff (No. NMW 90.9G3). Photo by the Photography Department, NMGW.

Plate 32.   Arborescent lycophyte cones. Left: *Lepidostrobus* from the South Wales Coal Measures at Abercarn, Gwent (x 1.3). Specimen in the collections of the National Museums and Galleries of Wales (No. G.1007). Right: *Sigillariostrobus* from Seam F (Westphalian) of Dortmund, Germany (x 1.1). Specimen in the collections of Museum fur Naturkunde, Berlin (No. 2690). Photo of *Lepidostrobus* by the Photography Department, NMGW. Photo of *Sigillariostrobus* by B. A. Thomas; printed by the Photography Department, NMGW.

Plate 33.   Model of a *Lepidostrobus* cone on a leafy shoot reconstructed from specimens such as the one figured in plate 32. The model was created by Annette Townsend and is now in the collections of the National Museums and Galleries of Wales, Cardiff (V97.26.5). Photo by the Photography Department, NMGW.

Plate 34.   Arborescent lycophytes are heterosporous producing both microspores and megaspores, sometimes together in one cone and sometimes in different cones. The upper microspores are called *Lycospora perforata* Bhardwaj and Venkatachala (x 2,000) and come from the microsporangiate cone *Lepidostrobus binneyanus* Arber. The lower megaspore is called *Lagenicula horrida* Zerndt (x 130) and comes from the bisporate cone *Flemingites gracilis* Carruthers. Scanning electron photos B. A. Thomas, printed by the Photography Department, NMGW.

Plate 35.   Certain arborescent lycophyte cones that have large sporophylls disintegrate on maturity. The sporophylls when found as adpressions are usually referred to *Lepidostrobophyllum*. If found permineralised they would probably be called *Lepidocarpon* or *Achlamydocarpon*. This transversely split cone from the Upper Westphalian D of Kilmersdon Colliery Tip, near Radstock, England contains sporophylls called *Lepidostrobophyllum alatum* Boulter. Specimen from the National Museums and Galleries of Wales, Cardiff (No. 90.8G). Photo by the Photography Department, NMGW. The cone base (x 1) and the two isolated sporophylls (x 1), from the same location, demonstrate the range of form found within the species. The smaller triangular sporophyll must have come from the apical region of a cone while the larger would have come from the middle region. Specimens from the Natural History Museum, London (Nos V.60430, V.60432). Photos by B. A. Thomas.

Plate 36.   A vertical section through a sporophyll of *Lepidocarpon* in a lower Westphalian coal ball from Lancashire. Specimen from the National Museums and Galleries of Wales, Cardiff (No. 98.24.G3). Photo. B. A. Thomas, printed by the Photography Department, NMGW.

Plate 37.   The detail of *Selaginellites gutbieri* (Göppert) Kidston is of the most complete and spectacular specimen of any fossil herbaceous lycophytes (x 4). First figured in 1885, from the Westphalian D of Oberhohdorf, near Oelsnitz in the Zwickau coalfield, Saxony, Germany, has a much dichotomising shoot with 26 terminal cones, up to 60 mm long. The spores from these heterosporous cones are most commonly found in those spore floras indicating a reduction in ground water and a partial elimination of many of the larger species of plants. This allowed the surviving plants to exploit the more open habitats. Herbaceous heterosporous lycopsids developed before the end of the Devonian. Specimen from the Richter Collection, Zwickau Museum Germany. Photo B. A. Thomas, printed by the Photography Department, NMGW.

Plate 38.   The type specimen of the early horsetail *Pseudobornia ursina* Nathorst from the Upper Devonian of Bear island, Arctic Ocean (x 1). It clearly shows foliage borne in whorls around the stem. The complex leaves are quite different from the foliage of more recent horsetails, which mostly tend to have simple leaves. Specimen from the Swedish Museum Natural History, Stockholm. Photo Y. Arremo, SMNH.

Plate 39.   The arborescent horsetail *Calamites* grew in the Carboniferous tropical swamp forests. The surface of its stems has a characteristic longitudinal ribbing, often together with circular scars where lower-order stems were attached. The distributional pattern of these scars is sometimes used to distinguish 'species', although how well these equate to the original whole-plant species concept is far from clear. This particular example of *Calamites multiramis* Weiss is from the upper Westphalian D of Kilmersdon Colliery, Radstock, Somerset (x 1.2). The specimen is from the collections of the National Museums and Galleries of Wales, Cardiff (No. 90.9G.1). Photo by the Photography Department, NMGW.

Plate 40.   Transverse section of the calamite stem *Arthropitys* from a section of coal ball from the Halifax Hard Coal, Langsettian, of Yorkshire, England. The ring of primary vascular bundles is surrounded by the radiating tracheids of the secondary wood. The cortical tissues have been lost and the pith cavity invaded by two stigmarian rootlets. It is in the collections of the National Museums and Galleries of Wales, Cardiff (No. 60.538.1). Photo by B. A. Thomas, printed by the Photography Department, NMGW.

Plate 41. The arborescent *Calamites* bore its leaves in whorls or rosettes around the stem, similar to the living horsetail, *Equisetum*. Unlike *Equisetum*, however, the whorls of leaves are not fused at their base, but are individually attached to the stem. This particular type of foliage with large, broad leaves, is known as *Annularia stellata* (Sternberg) Wood (x 1.2). The specimen originated from the upper Westphalian D of Lower Writhlington Colliery, near Radstock, Somerset, and is now in the collections of the National Museums and Galleries of Wales, Cardiff (No. 90.8G.4). Photo by the Photography Department, NMGW.

Plate 42. Model of a shoot with *Annularia* leaves, which would have been attached to a calamite horsetail growing in the Late Carboniferous tropical forests. The model was created by Annette Townsend and is now in the collections of the National Museums and Galleries of Wales, Cardiff (V97.26.15). Photo by the Photography Department, NMGW.

Plate 43. *Asterophyllites equisetiformis* Brongniart is another type of foliage borne by the arborescent horsetail *Calamites*, and can be distinguished from most species of *Annularia* by its much more slender leaves. The specimen originated from the upper Westphalian D of Lower Writhlington Colliery, near Radstock, Somerset, and is now in the collections of the National Museums and Galleries of Wales, Cardiff (No. 90.8G.3). Photo by the Photography Department, NMGW.

Plate 44. Cones were borne by the arborescent horsetail *Calamites* in clusters at the ends of vegetative branches. Left: *Pothocites grantoni* Paterson (x 2) is the cone of the Lower Carboniferous *Archaeocalamites*. This specimen is from the Glencartholm Volcanic Group (Holkerian-Asbian) of Glencartholm, Scotland and is now in the collections of the Natural History Museum, London (No. V.195). Photo from the Photographic Studio of the Natural History Museum. Right: *Palaeostachya* sp., a typical type of cone thought to have been borne on the calamite trees of the Late Carboniferous tropical coal-forming forests. This particular example is a from the Westphalian D Farringdon Formation, Lower Writhlington Colliery, Radstock, Somerset (x 2.5). The cone consists of alternating whorls of sporangia and sporophylls borne along the cone axis. Specimen from the National Museums and Galleries of Wales, Cardiff (No. 90.8G.8). Photo by the Photography Department, NMGW.

Plate 45. Model of a fertile shoot of a calamite horsetail with *Calamostachys* cones, as would have been found growing in the Late Carboniferous tropical forests. The model was created by Annette Townsend and is now in the collections of the National Museums and Galleries of Wales, Cardiff (V97.26.13). Photo by the Photography Department, NMGW.

Plate 46. *Calamostachys paniculata* Weiss (x 2.5) with relatively small cones was found in the upper Langsettian of Cattybrook Claypit near Bristol. It shows how the cones were borne in clusters on the calamite tree, although exactly where on the plant they were attached is not clear. The specimen is now in the collections of the National Museums and Galleries of Wales (No. 86.101G.54a). Photo by the Photography Department, NMGW.

Plate 47. Part of a large calamite cone, *Macrostachya infundibuliformis* (Brongniart) Schimper from the upper Westphalian D (Upper Carboniferous) of Kilmersdon, near

Radstock, Somerset (x 2). Most calamite cones tend to be small, delicate structures, but this one is much larger and looks superficially like a cone of one of the giant club mosses. However, this specimen clearly shows that the sporophylls are spirally arranged (albeit slightly skewed in this case due to distortion during fossilisation) and it is thus a horsetail cone. Specimen from the National Museums and Galleries of Wales, Cardiff (No. 90.9G.4). Photo by the Photography Department, NMGW.

Plate 48. Horsetails also grew in the southern high-latitudes during the Late Palaeozoic. This is part of a leaf-whorl of *Raniganjia bengalensis* (Feistmantel) Rigby from the Lower Permian of India (x 3). Although superficially similar to the tropical genus *Annularia*, *Raniganjia* has the leaves in the whorl fused to each other for over half of their length and is thus more similar to the living *Equisetum*. The specimen is now in the National Museums and Galleries of Wales, Cardiff (No. 77.25G.2). Photo by the Photography Department, NMGW.

Plate 49. Vertical section through an anatomically preserved example of a cone known as *Cheirostrobus pettycurensis* Scott (x 4). These cones are associated with stems that are very like the putative equisetalean *Sphenophyllum*. The sporangiophores are also very like those found in the cones borne by *Sphenophyllum*. However, they are far more complex than any of the more typical *Sphenophyllum*-borne cones, or of any other known equisetalean for that matter, consisting of large numbers of sporangia, as can be seen on the in the photograph. This specimen originated from the Visean Pettycur Limestone, near Burntisland, Fife, Scotland, and is now stored in the Natural History Museum, London (No. SC.3661). Photo by the Photographic Studio, Natural History Museum, London.

Plate 50. *Sphenophyllum* was a scrambling plant growing in the Late Carboniferous tropical swamp forests. It probably spread rapidly over newly-cleared parts of the river levees, resulting from storm-damage or bank-collapse. The leaves are formed in whorls, as with the horsetails. However, the individual leaves are wedge-shaped and have several radiating veins, and are thus quite different from more typical horsetails. This specimen of *Sphenophyllum cuneifolium* (Sternberg) Zeiller was from the middle Duckmantian of Rhigos, South Wales, and is now the Natural History Museum, London (No. V.63667). Photo by the Photography Department, NMGW.

Plate 51. A model of the *Sphenophyllum* as it would have appeared in life on the levees of the Late Carboniferous tropical forests. It was created by Annette Townsend and is now in the collections of the National Museums and Galleries of Wales, Cardiff (V97.26.5). Photo by the Photography Department, NMGW.

Plate 52. The shoot of a *Sphenophyllum* from the Cathaysian floral province. Plants of this type, which were probably ground-scramblers, occur abundantly in the Upper Carboniferous of Europe and North America, but die out in the lowermost Permian. In the Cathaysian floras, however, they persist through into the Upper Permian. This particular specimen is from the Lower Permian of Simugedong, Taiyuan East Hill, Shanxi. Photo by the Photography Department, NMGW.

Plate 53. Horsetails have a fossil history back to the Palaeozoic. By the Jurassic plants indistinguishable from the living *Equisetum* had evolved. *Equisetum* is sometimes divided into two subgenera on the basis of a number of characters; one of which is the presence or absence of tubers formed on the underground rhizomes. These specimens of stems and tubers of *Equisetum burchardii* (Dunker) Schimper (x 1.5) come from the Cretaceous Wealden beds (Barremian) near Hastings, Surrey. Specimens from the National Museums and Galleries of Wales, Cardiff (Nos 76.7G 110–118 & 157–161). Photo by the Photography Department, NMGW.

Plate 54.   Fertile shoots of the early fern-like *Calamophyton bicephalum* Leclerq and Andrews from the Middle Devonian (Eifelian) of Belgium. This species belongs to the Cladoxylopsida, a loose assemblage of plants with similar highly dissected vascular systems. The plant grew rhizomatously with dichotomising aerial shoots that bore sterile and fertile branches in low helices or pseudowhorls. The sporangia are borne on recurved branches. Specimen from the Palaeobotany Department, University of Liège (No. ULg 5011/609B]. Photo M. Fairon-Demaret, Liège.

Plate 55.   Some sterile foliage such as this *Alloiopteris quecifolia* (Brongniart) Potonié from the Lower Namurian of Cabrières, France can be related to genera of fertile ferns; in this case to *Corynepteris*. The division of the frond and some anatomical evidence suggests them to belong to the Zygopteridaceae. Specimen from the Palaeobotany collections of the University of Montpellier, France (x 1.6). Photo J. Galtier; printed by the Photography Department, NMGW.

Plate 56.   Transverse sections through two permineralised rachises of the zygopterid fern *Metaclepsydropsis duplex* (Williamson) Bertrand, from the Pettycur Limestone (Asbian) of Pettycur, Scotland (x 6). They are parts of the upright frond of a scrambling plant. In the lower rachis, the hour-glass vascular trace has given off a curved trace leading to a pinna. The vascular trace in the upper rachis has also given off a leaf trace, but here it has divided into two strands. The thin section is in the collections of the National Museums and Galleries of Wales, Cardiff (No. 11.64.2). Photo by the Photography Department, NMGW.

Plate 57.   Another transverse section through a permineralised rachis of the zygopterid fern *Metaclepsydropsis duplex* (Williamson) Bertrand, from the Pettycur Limestone (Asbian) of Pettycur, Scotland (x 10). It shows two pinnae as they depart from the main rachis, each having its distinctive 'C'-shaped vascular strand. A third, completely separated pinna is also visible, presumably having been emitted from lower down the rachis. The thin section is in the collections of the National Museums and Galleries of Wales, Cardiff (No. 60.538.6). Photo by the Photography Department, NMGW.

Plate 58.   Another type of early fern found in the Carboniferous is *Stauropteris*, known exclusively from petrified fossils such as this one (*S. oldhamia* Binney) from the lower Langsettian of northern England (x 60). This transverse section through the stem shows the four-lobed vascular strand that is characteristic for this genus. The microscope slide from which this photograph was taken is in the collections of the National Museums and Galleries of Wales, Cardiff (No. 98.24.G4). Photo by B. A. Thomas. Printed by the Photography Department NMGW.

Plate 59.   *Cyathocarpus arborescens* (Brongniart) Weiss from the Coal Measures of Chaumoune Canton, Switzerland (x 1). Specimen from the Natural History Museum, London (WC.223 and GC.789). Photo by the Photographic Studio, Natural History Museum, printed by the Photography Department NMGW.

Plate 60. While many fern fossils are the remains of sterile foliage, fertile specimens occur reasonably commonly. This example is a fragment of the Late Carboniferous fern *Lobatopteris miltoni* (Artis) Wagner, showing pinnules with two rows of sporangial clusters (known as synangia) on either side of the mid-vein, a feature which clearly places it in the extant family the Marattiaceae. However, in contrast to the living marattiaceans, each synangium in this fossil species usually consists of only four sporangia. The upper photograph shows the whole specimen magnified x 1.5; the lower photograph shows a close-up of the same specimen coated with ammonium hydroxide to enhance the surface detail, x 3. The specimen, which originated from the middle Westphalian of Clay Cross, Derbyshire, and is now in the collections of the British Geological Survey, Keyworth (No. 5073). Photo by the Photography Department, NMGW.

Plate 61. Croziers of a marattialean fern from the Middle Westphalian of the British Coal Measures at Sandwell Park, near Dudley, West Midlands. Specimen from the Natural History Museum London (No. V.1265). Photo by the Photographic Studio of The Natural History Museum.

Plate 62. The trunks of the marattialean tree ferns of the Upper Carboniferous tropical forests bore distinctive scars in their upper regions, where fronds have been shed as the tree grew. The distribution and the shape of the scars would vary between species, in this case occurring as a double row of round scars. The distinctive trace of the vascular tissue of the frond petiole can be seen in the centre of each scar. This particular type of scar is known as *Caulopteris anglica* Kidston and is from the Westphalian D of Radstock, Avon (x 1.5). Specimen from the Natural History Museum, London (No. V.3088). Photo by the Photography Department, NMGW.

Plate 63. When preserved as petrifactions, the stems of the Late Palaeozoic marattialean ferns are known as *Psaronius*. This thin section, from the late Stephanian silicified peat of Autun, France, shows a transverse section through the outer part of such a stem with the numerous adventitious roots (x 4). Especially in the lower part of the tree, this zone of rooting tissue provided the bulk of the trunk. The thin section is in the collections of the National Museums and Galleries of Wales, Cardiff (No. 11.108.84). Photo by the Photography Department, NMGW.

Plate 64. A second major group of tree ferns growing in the Late Carboniferous tropical forests are referred to the now extinct family, the Tedeleaceae. This is part of the frond of one of the most widely found species, *Pecopteris plumosa* (Artis) Brongniart, showing its distinctive, small, tooth-like pinnules (x 3). Unlike the marattialean ferns, with which they are often associated, the fertile pinnules of this family have the sporangia near the margins. The specimen was found in Duckmantian aged strata at Warren House Opencast near Wakesfield, Yorkshire, and is now in the National Museums and Galleries of Wales, Cardiff (No. 92.20G.1). Photo by the Photography Department, NMGW.

Plate 65. Tree fern foliage represents the most commonly found fern-fossils in the Upper Carboniferous. However, there are also the remains of smaller ferns, probably representing the understorey vegetation growing on the levees within the forests. This example from the Westphalian of the Saarland Coalfield, Germany, has very characteristic, dissected pinnules and can be assigned to the species *Zeilleria frenzlii* (Stur) Gothan (x 2.5). The sporangia found at the tips of the pinnule lobes suggest that *Zeilleria* is a eusporangiate fern.

However, its relationship to the living ferns is otherwise uncertain and some palaeobotanists now assign it to an extinct order, the Urnatopteridales. The specimen is now in the collections of the Saarbrücken Mining School (No. C/746). Photo C. Cleal, printed by the Photography Department, NMGW.

Plate 66. A polished section of a Permian osmundaceous tree fern, *Palaeosmunda williamsii* Gould (x 4), from Coleron, nr Rockhampton, Queensland, Australia. The stem with its surrounding mantle of leaf bases is preserved in chalcedony. Specimen from the Natural History Museum London (V.5000). Photo from the Photographic Studio of the Natural History Museum.

Plate 67. Fragments of two osmundaceous ferns from the Middle Jurassic of Yorkshire. Main Photo: *Cladophlebis denticulata* (Brongniart) Nathorst, from Cayton Bay (x 2). This shows a part of a sterile osmundaceous frond, typical of those found in the Jurassic. Specimen from the National Museums and Galleries of Wales (No. 98.24.G5). Photo by the Photography Department, NMGW. Bottom left: *Osmundopsis hillii* van Konijnenburg-van Cittert from Hasty Bank (x 3). This is part of a fertile frond showing the reduced pinnules bearing numerous sporangia. Specimen from the Natural History Museum, London (No. V.60953). Photo by the Photographic Studio of the Natural History Museum.

Plate 68. Whole plants are very rarely preserved as fossils. This almost complete juvenile fern, *Arnophyton kuessii* Ash and Tidwell, came from the Lower Permian of New Mexico (x 2.7). It has a horizontal rhizome with roots, a short thick upright aerial stem with a tuft of leaves has both juvenile and more dissected adult fronds. Specimen from the New Mexico Museum of Natural History, Alberquerque (No. D31-P1056). Photo S. R. Ash, Alberquerque, printed by the Photography Department, NMGW.

Plate 69. *Weichselia reticulata* (Stokes & Webb) Fontaine was a particularly successful member of the Matoniaceae forming dense thickets in an analogous way to living bracken. It possessed a number of xeromorphic features such as a cuticle, fibrous tissues and sunken stomata suggesting that it grew in areas that were subjected to periods of drought. This specimen from the Wealden (Cretaceous) of Ecclesbourne, East Sussex shows many charcoalified frond fragments representing the remains of ferns burnt by spontaneous fire (x 3). Specimen from the Natural History Museum, London (No. V.708). Photo from the Photographic Studio, Natural History Museum, London, printed by the Photography Department, NMGW.

Plate 70. *Lygodium kaulfussi* Heer has been described from the Miocene, Eocene and Oligocene of Europe, the Eocene of several States in the USA and possibly the Eocene of China. Similar species have been found in the Eocene of South America and Australia. The present specimens are part of a large collection from the Upper Eocene of Wyoming, USA. The branches are non-laminate with fertile elongated clusters of sporangia at the ends of a palmately organised dichotomising branching system (x 6). All the sterile, palmately lobed, leaflets (bottom left, x 1.5) have small a semicircular scar at their base indicating that they were abscised at the junction between the leaflet and its stalk. Some, but not all, extant species loose their leaflets in this way. *Lygodium* belongs to the extant family the Schizaeaceae, which has a fossil record back to the Jurassic although there are ferns that show possible affinity back to the Carboniferous. Specimen from the Florida Museum of

Natural History, Gainsville (Nos 5307 and 5304). Photo S. R. Manchester, Florida Museum of Natural History, Gainsville; Printed by the Photography Department, NMGW.

Plate 71.   This matoniaceous fern *Phlebopteris smithii* (Daugherty) Arnold is from the Chinle Formation (Triassic) of the Petrified Forrest National Park, New Mexico (x 2). The typical palmate arrangement of the frond is clearly visible. The only living genus of the Mantoniaceae, *Matonia*, is restricted to Southeast Asia. Specimen from the US National Museum, Washington, DC (No. 298829A). Photo S. R. Ash, Albuquerque, printed by the Photography Department, NMGW.

Plate 72.   There are several lengths of the floating heterosporous fern *Azolla primaeva* Dawson with its long dangling roots on this specimen from the middle Eocene of Princeton, British Columbia, Canada (x 4). Specimen from the University of California at Berkeley, Museum of Paleontology (No. 94720). Photo M. E. Collinson, London.

Plate 73.   This megaspore, called *Arcellites hexapartitus* (Dijkstra) Potter comes from the Wessex Formation (Lower Cretaceous/Barremian) of the Isle of Wight (x 100). It has prominent flanges to the lips of its triradiate mark, long appendages and many long, hairlike processes extending from the surface of the body wall. This megaspore most probably belonged to an extinct group of 'water ferns' related to the extant families Marsileaceae (*Marsilea*, *Pilularia*) and Salviniaceae (*Salvinia*, *Azolla*). Photo D. J. Batten; printed by the Photography Department, NMGW.

Plate 74.   Fragments of the leaf *Sphenopteridium rigidum* (Ludwig) Potonié from the Upper Devonian of Plaistow Quarry, north Devon (x 3.5). These small fragments were almost certainly parts of the foliage of very early seed plants belonging to the Elkinsiales and, at other localities, have been found associated with small cupulate ovules. The specimen is now stored in the Natural History Museum, London (No. V.3562). Photo by the Photographic Studio, Natural History Museum, London.

Plate 75.   Vertical section through an anatomically preserved ovule from an Early Carboniferous gymnosperm (x 75). Such ovules are thought to have been borne by various different types of early gymnosperm and, when found isolated, are referred to as *Hydrasperma tenuis* Long. This specimen clearly shows that the preintegument forms an open collar around the top of the seed, rather than enclosing it with just a small opening (micropyle) as in modern gymnosperms. Also visible is the extension of the integument to form a structure known as a lagenostome, which assisted in pollen-capture. The specimen originated from the Visean of Oxroad Bay, East Lothian, Scotland, and is now stored in the Royal Museum of Scotland, Edinburgh (No. OBC084hT/48). Photo Richard Bateman, Royal Botanic Gardens, Edinburgh.

Plate 76.   These petrified plant fossils from Tournaisian (Early Carboniferous) ash deposits at Oxroad Bay, Lothian, Scotland, show the fine anatomical detail that can often be preserved in this type of deposit. Upper left is a cross-section through the stem of a shrubby pteridosperm, *Tetrastichia bupatides* Gordon, with its characteristic four-lobed stele (x 10). Specimen now in the Gordon Collection, the Natural History Museum, London. Upper right is a section through the stem of a small club-moss *Oxroadia gracilis* Alvin (x 140). Specimen from the Bateman Collection (No. OBD(?2.15)038bT/2). Lower left shows two pteridosperm ovules of the type known as *Hydrasperma tenuis* Long when found isolated, but here seen attached to a cupule known as *Pullaritheca longii* Rothwell and Wight (x 45). Specimen in the Hancock Museum, Newcastle-upon-Tyne (No. 11718). Lower right shows sporangia from the cone of the sphenophyte *Protocalamostachys farringtonii* Bateman, the lower one containing microspores, the upper one megaspores (x 85). Specimen from the Bateman Collection (No. OBC084gB/5). Upper left photo G. W. Rothwell, others R. M. Bateman.

Plate 77. A sterile frond of the lagenostomalean pteridosperm *Sphenopteridium pachyrrachis* (Göppert) Schimper, showing the main dichotomy of the primary rachis (natural size). This specimen came from the Visean Series of Glencartholm, southern Scotland. Specimen from the Natural History Museum, London (No. V.186). Photo by the Photographic Studio, Natural History Museum, London.

Plate 78. A sterile frond of another Early Carboniferous lagenostomalean pteridosperm, *Adiantites machanekii* Stur (natural size). This shows basically the same frond architecture as the previous specimens, but has much more slender, feathery pinnules. The specimen originated from the Visean of Teilia Quarry, near Prestatyn, North Wales, and is now stored in the Natural History Museum, London (No. V.2755). Photo by the Photographic Studio, Natural History Museum, London.

Plate 79. Terminal part of the stem of *Diplopteridium holdenii* Lele and Walton, showing numerous fronds in attachment (x 1.1). At least one of the fronds in the terminal cluster (that on the left hand side) shows clusters of ovules attached in the fork of the main dichotomy of the primary rachis. Also visible in the lower left-hand side of the specimen (upside down as viewed) is the terminal part of a shoot showing young fronds still enrolled in a fern-like crozier. The specimen came from the Lower Carboniferous (Visean) Drybrook Sandstone of the Forest of Dean, and is now stored in the Natural History Museum, London (No. V.62331b). Photo N. P. Rowe, Université des Sciences et Techniques du Languedoc, Montpellier.

Plate 80. This specimen, shown at natural size, represents an almost complete frond of the lagenostomalean pteridosperm *Mariopteris sauveurii* (Brongniart) Frech. Most pteridosperm remains found in the Carboniferous were from shrubs and small trees, but this is the leaf of a vine-like plant that climbed the trunks of trees to reach the upper reaches of the forest canopy. This specimen was found above Seam 26, Sulzbach Formation (Bolsovian) at the Frankenholz Mine, Saarland, Germany. It is now in the collections of the Saarbrücken Mining School (No. B/305). Photo C. Cleal, printed by the Photography Department, NMGW.

Plate 81. Part of a frond of *Mariopteris nervosa* (Brongniart) Zeiller, a climbing plant from the Latre Carboniferous tropical forests of Europe (x 2). It clearly shows the elongate tip to each pinna, which may either be a drip-tip to help excess rainwater to run off the leaf, or may be clasping structures to help it climb round trees. The specimen is from an unknown locality in the British Coal Measures, and is now stored in the National Museums and Galleries of Wales, Cardiff (No. 30.232G.165). Photo by the Photography Department, NMGW.

Plate 82. Another type of lagenostomalean pteridosperm frond commonly found in the Upper Carboniferous is *Eusphenopteris*. Like most pteridosperms, this originally consisted of a forking frond, the illustrated specimen (of *E. sauveurii* (Crépin) Simson-Scharold, x 1) being one of the major branches produced by the fork. It was probably the leaf of a small shrub. The specimen came from Bolsovian strata at Helene Mine, Saarland, Germany. It is now in the collections of the Saarbrücken Mining School (No. C/142). Photo C. Cleal, printed by the Photography Department, NMGW.

Plate 83. Medullosalean fronds were anything up to 7 m long, with numerous small, usually elongated pinnules. This plate shows an examples of part of a *Neuralethopteris jongmansii* Laveine frond, which is a species characteristic of the early phases in the development of the Late Carboniferous tropical forests (x 1.25). The specimen came from the lower Langsettian of Nant Llech, South Wales, and is now stored in the Natural History Museum (No. V.23359). Photo by the Photographic Studio, Natural History Museum, London.

Plate 84. Segment of a large frond known as *Alethopteris serlii* (Brongniart) Göppert from the upper Westphalian D (Upper Carboniferous) of Kilmersdon, near Radstock, Somerset (x 2). The entire fronds of these medullosalean pteridosperms could be anything up to 7 m long, but mostly we find much smaller segments of the entire frond, such as this. When first found in the early years of the nineteenth century, they were thought to be the remains of the fern *Polypodium*, to which they bear a passing resemblance. However, we now know that they are foliage of a seed plant and very occasionally seeds or ovules may be preserved still attached to the frond. The specimen is stored in the National Museums and Galleries of Wales, Cardiff (No. 35.594G.42). Photo by the Photography Department, NMGW.

Plate 85. *Trigonocarpus* is the seed borne by the medullosalean pteridosperms, which is sometimes preserved as a compression as shown in the main part of the plate (x 3.5). This clearly shows the outer, fleshy part of the seed and the central 'nut'. More commonly found, however, are casts of the 'nut', such as shown in the inset, where the characteristic three longitudinal ribs are preserved (x 2). The main specimen, which came from the Westphalian of Newcastle-upon-Tyne, is now in the Natural History Museum, London (No. V.40584). The 'nut' is from the Duckmantian Peel Hall Rock, near Bolton, and is in the collections of the National Museums and Galleries of Wales, Cardiff (No. 24.43G.107a). Photos by the Photography Department, NMGW.

Plate 86. This thin section shows a transverse section through the medullosalean seed *Trigonocarpus parkinsonii* Brongniart (x 5). It occurred in a coal ball from the lower Langsettian Lower Foot Seam at Shore, Lancashire, one of the classic sites for Upper Carboniferous petrifactions. The section clearly shows the three ribs that ran longitudinally along the hard sclerotesta of the seed and which can be seen in the 'nut' figured in the previous plate. Also visible is the fleshy outer tissue of the seed (sarcotesta) and the nucellus in the centre. The thin section is in the collections of the National Museums and Galleries of Wales, Cardiff (No. 21.77.54). Photo by the Photography Department, NMGW.

Plate 87. Pteridosperm fronds are normally preserved as relatively small fragments, which makes it difficult to work out the branching of the rachises. Occasionally, however, larger fragments have been unearthed, which have allowed us to reconstruct the fronds more fully. This is an example of part of a *Neuropteris ovata* Hoffmann frond, shown x 0.25, found in a large sandstone block at Point Aconi, on the coast of Cape Breton Island, Canada. It is thought that the sandstone was part of a channel deposit formed near the levee

Explanations of plates

bank on which the tree had been growing, and that the frond had not been subjected to any significant transportation before being entombed in the sediment. The specimen is now in the collections of the University College of Cape Breton, Sydney, Nova Scotia (No. 985GF-248). Photo E. L. Zodrow, Sydney NS.

Plate 88.   Cuticle prepared from the leaf of the Late Carboniferous medullosalean pteridosperm *Neuropteris ovata* Hoffmann from the Emery Seam, Glace Bay, Sydney Coalfield, Cape Breton (x 500). The photograph was taken using interference phase contrast illumination, to enhance the epidermal cell pattern preserved on the cuticle. This example came from the edge of one of the pinnules, and shows cuticles from both the upper (to the left) and lower (to the right) surfaces. The upper cuticle shows characteristically (for this species) sinuous cell walls. The lower cuticle shows epidermal cells with straighter walls, but with stomata and the bases of two hairs. Also on the lower cuticle, near the very edge of the pinnule, are numerous round structures with holes in the middle, which are the positions of hydathodes. These are pores from which the plant can exude excess water (a process known as guttation) and feature commonly in plants of tropical rain forests. The cuticle is mounted on a microscope slide the collections of the University College of Cape Breton, Sydney, Nova Scotia (No. 985GF-248). Photo C. Cleal and printed by the Photography Department, NMGW.

Plate 89.   Part of a *Paripteris gigantea* (Sternberg) Gothan frond, showing pinnules attached to two orders of rachises within the frond (x 2). It belongs to the parispermacean pteridosperms, which first arose in the Early Carboniferous of China and then spread out over much of the tropical belt during the Late Carboniferous. The specimen originated from strata of the Faisceau de Meunière (Middle Westphalian) in the Dechy Mine, Douai, northern France. Specimen from the collections of the Laboratory for Palaeobotany, University of Lille, France (No. 947). Photo J.-P. Laveine, Lille.

Plate 90.   *Linopteris subbrongniartii* (Grand'Eury) Fritel. This is another type of parispermacean frond, distinguished from *Paripteris* by the mesh veining of the pinnules. It is assumed that *Linopteris* evolved from *Paripteris*, although there is no direct evidence of a transition between the two leaf-types. The upper photograph shows a large part of a frond, probably from near the apex, showing pinnules attached to all orders of rachis (x 0.25). Specimen from the Saint Albert Seam (middle Westphalian), Pit 3 W, Méricourt, northern France; now stored in the collections of the Lewarde Centre Historique Minier (No. CHM 1). The lower specimen shows a close up of some of the pinnules showing the characteristic mes veining (x 3). Specimen from the Saint-Jean-Baptiste Seam (Middle Westphalian), Pit No. 2, Béthune, northern France; now stored in the collections of the Laboratory for Palaeobotany, University of Lille, France (No. 947). Photos J.-P. Laveine, Lille.

Plate 91.   Most plant fossils represent fully developed organs, but occasionally we find juvenile structures formed that can provide valuable evidence of growth patterns. This specimen (x 1), from the Middle Westphalian of Ebbw Vale, South Wales, shows a young frond of the potonieaceaen pteridosperm *Paripteris pseudogigantea* (Potonié) Gothan, which had not fully unrolled when it was fossilized. The frond appears to have started as an enrolled crozier, as in most ferns, suggesting that this was a primitive mode of development shared by several independent lineages of plants. Also of interest are the prominent spines of the main rachises, which may have been to protect the young croziers, which would have been particularly vulnerable to predation. The specimen is stored in the Natural History Museum, London (No. 41283). Photo by the Photographic Studio, Natural History Museum, London.

Plate 92. Associated with *Paripteris* and *Linopteris* fronds are distinctive pollen-bearing organs known as *Potoniea*, after which the family is named. Single pollen organs are most normally found, but occasionally clusters of them are found, as in this example from the Duckmantian of Dudley, West Midlands. Each of the cup-like structures consists of clusters of elongate pollen-bearing tubes. This specimen is stored in the Natural History Museum, London (No. V.33301). Photo by the Photography Department, NMGW.

Plate 93. A segment of the peltasperm frond *Autunia conferta* (Sternberg) Kerp from the Lower Permian Lauterecken-Odernheim Formation, Lebach or Berschweiler, Germany (x 1.4). The specimen shows the characteristic intercalated pinnules on the primary rachis. Pteridosperms of this type rapidly developed a world-wide distribution during the Permian and were able to survive the Permian-Triassic extinction event that saw the demise of many other groups of Palaeozoic gymnosperms. Specimen from the Weiss Collection, Museum für Naturkunde, Berlin (No. 59/9/1). Photo H. Kerp, Westfälische Wilhelms-Universität, Münster.

Plate 94. A small frond of the peltasperm *Supaia shanxiensis* Wang from the Lower Permian Tianlongsi Formation, Wangtao village, Qinyuan District, Shanxi, China (x 3). This type of foliage was originally described from the Permian of North America, but specimens such as this are now known to occur also in North China. Although the fronds look very different from those of *Autunia*, which is common in the Lower Permian of Europe (see Pl. 93), the associated seed-bearing cones are very similar and shows that they both belong to the same family of plants, the Peltaspermaceae. Specimen from the Institute of Geology and Mineral Resources, Tianjin, Shanxi, China (No. 9306–2). Photo Wang Ziqiang, IGMR, Tianjin.

Plate 95. The leaf of the gigantopteroid *Gigantonoclea crenata* Wang from the Lower Permian of Nanhegou village, Baode District, Shanxi, China (x 6). These types of leaves are often associated with peltasperm foliage (such as shown in the previous two figures) in the Permian of China and North America. These entire types of leaf superficially resemble angiosperm foliage but were in fact borne on gymnospermous plants. Specimen from the Institute of Geology and Mineral Resources, Tianjin, Shanxi, China (No. 9075–3). Photo Wang Ziqiang, IGMR, Tianjin.

Plate 96. Another type of gigantopteroid leaf, this time of *Gigantonoclea lagrelii* (Halle) Koidzumi, from the Lower Permian of Huoshan village, Jiacheng District, Shanxi, China (x 2.5). As can be seen, this was a segmented leaf that was significantly larger than the previous species (although still much smaller than the pteridospermous leaves found in the Carboniferous). Specimen from the Institute of Geology and Mineral Resources, Tianjin, Shanxi, China (No. TH7126a–87). Photo Wang Ziqiang, IGMR, Tianjin.

Plate 97. Foliage of the glossopteridalean tree *Lanceolatus lerouxides* Plumstead from the Vryheid Formation (Middle Ecca, Lower Permian), Leeukuil quarries, Vereeniging, South Africa. These plants dominated much of the Permian vegetation of the southern middle and high latitudes (known as Gondwanaland). Leaves of this type where the reproductive structures are unknown tend to be placed in the form-genus *Glossopteris*. In this case, however, reproductive structures are known, as seen in the left-hand figure (x 1.3). The elongate, ovule-bearing capitum can be seen in the middle of the leaf and is thought to be attached to its midvein. The right-hand figure shows a close-up of a leaf, demonstrating the characteristic mesh-veining of this genus (x 4.2). Left-hand specimen from the Bernard Price Palaeontological Institute, Johannesburg (No. BP/2/-); right-hand specimen from the Vereeniging Museum, Transvaal (No. VM/03/3205/22). Photo H. Anderson, National Botanical Institute, Pretoria.

Plate 98. Sterile glossopteridalean foliage from the Estcourt Formation (Upper Permian), Mooi River, National Road, Natal, South Africa. Although both of these specimens are sterile, fertile specimens have been reported that prove that both species figured here belong to the genus *Lidgettonia*. The plate shows two of the characteristic species of the Estcourt Formatrion of the South African Karoo Basin, *Lidgettonia mooiriverensis* Anderson and Anderson (left, x 2) and *L. lidgettonioides* (Lacey et al.) Anderson and Anderson (right, x 1.75), distinguishable by the shape of the leaves. Note the quite different style of venation to that shown in the previous figure. Specimens from the Bernard Price Institute for Palaeontology, Johannesburg (Nos BP/2/7563 and BP/2/8024, left and right, respectively). Photo H. Anderson, National Botanical Institute, Pretoria.

Plate 99. The cordaites were a group of Carboniferous and Permian plants that were closely related to the early conifers. Left: The leaves, *Cordaites*, were long and strap-shaped with many parallel veins (x 1). Right: The reproductive organs were complex structures, such as this seed-bearing 'cone' (*Cordaitanthus* sp.) found in a middle Westphalian ironstone nodule, Phoenix Brickworks, Crawcrook, Durham (x 1.6). This is in fact not a simple cone but a compound structure. A series of small leaves are borne on either side of the main stem and in the angle of each there is attached a small cone. Each cone consists of a series of bracts and sporangia- or (as in this case) ovule-bearing axes. In the illustration, the ovules can be clearly seen in the upper part of the specimen, on the left-hand side. Specimens from the Natural History Museum, London (Nos V.29275 and V.28321). Photo by the Photography Department, NMGW.

Plate 100. A leafy shoot of an early conifer (*Culmitzschia laxifolia* (Florin) Clement-Westerhof) from the Lower Permian Nahe Group, Sobernheim, Germany (x 2). Note the slender branches with widely-spaces, rigid leaves. Although the foliage is remarkably 'modern' in appearance, the cones of these plants were very different from those of today's conifers and resemble rather more the inflorescences of the Late Carboniferous cordaites. Also present is a fragment of marattialean fern frond. Specimen from the collection of the Laboratory of Palaeobotany and Palynology, Utrecht (No. 13145). Photo H. Kerp, Westfälische Wilhelms-Universität, Münster.

Plate 101. Another leafy shoot of an early conifer (*Culmitschia parvifolia* (Florin) Kerp and Clement-Westerhof) from the Lower Permian Lauterecken-Odernheim Formation, Oderheim, Germany (x 2). This has wider branches than the previous specimen, but with more delicate leaves. Specimen from the Paläontologisches Institut, Universität Mainz, Germany (No. Q 1569). Photo H. Kerp, Westfälische Wilhelms-Universität, Münster.

Plate 102. Pieces of fossilised wood are commonly in the Mesozoic and Tertiary fossil record, this particular silicified example came from the Lower Cretaceous near Sevenoaks,

Kent (x 0.6). Its woody texture, including some knots, are clearly visible, but without evidence of the structure of the wood (e.g. next plate), it can be difficult to be sure of the type of tree that it originated from. However, in view of its age and the types of foliage normally found in rocks of this age, it is most likely to be coniferous. The specimen is now in the National Museums and Galleries of Wales, Cardiff (No. 50.140G.1). Photo by the NMGW Photo Studio.

Plate 103.   Triassic petrified wood was used by American Indians, in what is now the south-western United States, for arrowheads and other implements. Some also used large blocks as building stones for their houses. This small pueblo, now called 'Agate House', made of sections of petrified logs is in the Petrified Forest National Park, Arizona. It was built and occupied by American Indians about 600 years ago, although recently it has been partially restored. The Navajo have a tradition that the petrified logs are bones of the Great Giant, or Monster, Yietso whom their forefathers killed when they occupied the country. The Pinutes of Utah, however, might have recognised that the material resembled wood even though they believed the logs to be shafts of their thunder god Shinauav. Photo S. R. Ash, Albuquerque, printed by the Photography Department, NMGW.

Plate 104.   While conifer foliage occurs widely in terrestrial Mesozoic sediments, cones tend to be much less common and cones attached to leafy shoots rarer again. This example, from the Middle Jurassic Gristhorpe Plant bed at Cayton Bay, Yorkshire, shows a female cone attached to a shoot of *Elatides williamsonii* (Brongniart) Nathorst (x 1.5), and belongs to the Taxodiaceae. The specimen is in the collections of the National Museums and Galleries of Wales, Cardiff (No. 84.27G.361). Photo by the Photography Department, NMGW.

Plate 105.   *Ginkgoites pluripartita* (Schimper) Seward, a ginkgo-like leaf from the Lower Cretaceous (Wealden) of Osterwald, Germany (x 2.2). Leaves such as this, which show a general similarity to leaves of the living *Ginkgo biloba* L., occur commonly in the Mesozoic. Reproductive structures are much rarer, however, and so many palaeobotanists place such leaves in a separate form-genus (*Ginkgoites*) to make it clear that we are not absolutely sure of their affinities. This specimen is in the collections of the Geologisch-Paläontologisches Institut und Museum der Georg-August-Universität, Göttingen, Germany (No. P4–5). Photo by J. Watson, Manchester, printed by the Photography Department, NMGW.

Plate 106.   *Crossozamia chinensis* Gao and Thomas, the ovule-bearing sporophyll of an early cycad cone (x 4). The sporophyll has still retained some of the ovules in attachment to its lower part. Although nearly 300 million years old, these are remarkably similar to the sporophylls in living *Cycas* cones, indicating that the Cycadaceae is the most primitive of the cycad families. Specimen from the Lower Permian of Simugedong, Taiyuan East Hill, Shanxi, China. The specimen is in the collections of the National Museums and Galleries of Wales (No. 98.24.G7). Photo by the Photography Department, NMGW.

Plate 107.   The cycad leaves, *Nilsonia*, are relatively abundant in Jurassic strata. This specimen shows two species of that genus, *N. tenuinervis* Seward with the undivided or only shallowly divided leaves, and *N. compta* (Phillips) Bronn with the more deeply-divided leaves and coarser venation (x 1.25). The specimen originated from the classic Middle Jurassic Gristhorpe Plant Bed at Cayton Bay, Yorkshire, from where similar material has been shown to yield cuticles confirming that these are true cycads. Also found here are both male and female cones which, based on their association together at several other Yorkshire Jurassic sites, confirm that these are cycad leaves. The specimen is now in the National Museums and Galleries of Wales, Cardiff (No. 76.7G.215). Photo by the Photography Department, NMGW.

Plate 108. A reconstruction of a leafy shoot of the Middle Jurassic cycad known informally as the *Beania* tree. It was created by Annette Townsend and was based on the work of Tom Harris on the Yorkshire Jurassic floras. The model is now in the collections of the National Museums and Galleries of Wales, Cardiff (No. V97.26.8). Photo by the Photography Department, NMGW.

Plate 109. Bennettite leaves are very similar to cycad foliage and can be difficult to distinguish unless there are cuticles or associated reproductive structures. This example, from the Upper Triassic Lunz Lettenkohle near Linz, Austria, shows part of a leaf of one of the oldest known bennettites: *Pterophyllum approximatum* Stur (x 1.25). It not only yields the distinctive cuticles of the bennettites, but is closely associated with reproductive structures of that group. The specimen is now in the National Museums and Galleries of Wales, Cardiff (No. 76.7G.188). Photo by the Photography Department, NMGW.

Plate 110. This example of *Otozamites bechei* Brongniart is another type of Triassic bennettite leaf, this time from the Rhaetic of Charlton Mackrell, Somerset (x 1). Plant fossils are extremely scarce in the British Rhaetic and this represents by far the most complete and best preserved example of a bennettite leaf found from these deposits. The specimen is now in the National Museums and Galleries of Wales, Cardiff (No. 90.5G.3). Photo by the Photography Department, NMGW.

Plate 111. Cuticles from the frond of the bennettite *Nilssoniopteris vittata* (Brongniart) Florin, from the Middle Jurassic of Yorkshire (x 400). Both the upper (upper photo) and lower (lower photo) cuticles show epidermal cells with strongly sinuous walls. The lower cuticle also has stomata with very prominent guard cells, which are characteristic of the bennettites. The specimen is now in the National Museums and Galleries of Wales, Cardiff (No. WSL.174). Photo by C. Cleal, printed by the Photography Department, NMGW.

Plate 112. This is a vertical section through a bennettite 'flower' *Cycadeoidea*, from Cycad Valley, eastern Black Hills, South Dakota, USA (x 3). It shows the central receptacle, which bore the ovules and, at its distal end, the remains of pollen sacs. The entire structure is surrounds by numerous protective bracts. This example appears to have been a young 'flower' which had not fully opened. The specimen is now in the National Museums and Galleries of Wales, Cardiff (No. 21.146.8). Photo by the Photography Department, NMGW.

Plate 113. The Jurassic fruit-bearing organ *Caytonia sewardii* Thomas (x 40 with an enlargement) comes from the famous Gristhope Bed (Bajocian) at Cayton Bay, Yorkshire. The fruits were edible, like berries, for their coprolites have been found containing their chewed up remains. They contained numerous ovate seeds. The mouth with its lip curving up to the left is itself to the left of the broken stalk. Scanning Electron Photo B. A. Thomas, printed by the Photography Department, NMGW.

Plate 114. A more or less entire frond of the corystosperm *Dicroidium dubium* (Feistmantel) Gothan from the Molteno Formation (Upper Triassic) of South Africa (x 1.7). It clearly shows the dichotomy of the main rachis near the bottom of the frond. This was one of the dominant types of plant in the Late Triassic of Gondwana. Specimen from the National Botanical Institute, Pretoria, South Africa (No. PRE/F/170). Photo H. Anderson, NBI, South Africa.

Plate 115.   These are the upper and lower cuticles of a pinnule of the Jurassic corystosperm leaf *Pachypteris lanceolata* Brongniart (x 20). The specimen is from the Bathonian of Roseberry Topping, Yorkshire and is now in the collections of the National Museums and Galleries of Wales, Cardiff (98.24.G6). Photo B. A. Thomas, printed by the Photography Department, NMGW.

Plate 116.   A cut slab of a silicified palm from the Cretaceous of Austin, Texas, USA (x 3). Palms, being monocotyledonous angiosperms, have many individual vascular bundles in their stems and do not form secondary wood as do conifers and dichotyledenous angiosperms. Specimen from the National Museums and Galleries of Wales, Cardiff (No. 78.55G.1). Photo by the Photography Department, NMGW.

Plate 117.   This is an unnamed angiosperm fruit from the Dakota Formation, Middle Cretaceous, of central Kansas, USA (x 2). The follicles are borne helically in an ovoidal head showing it to be most similar to Recent magnoliid angiosperms. The specimen is in the collections of the Florida Museum of Natural History, Gainsville (No. 15706–3084). Photo D. Dilcher, printed by the Photography Department, NMGW.

Plate 118.   Fossil twigs with attached leaves and reproductive organs are very rare, but are more common in some groups than others. There is the phenomenon of 'self-pruning' in *Ulmus* (elms) and *Populus* (poplars) which improves the chances of twigs becoming fossilised. This specimen of *Cedrelospermum lineatum* (Lesquereux) Manchester (x 2.5) from the early Oligocene Florissant beds of central Colorado in the United States can be referred to the extant family of elms, the Ulmaceae. The order of plants including the elms, the Ulmoideae, evolved and diversified during the latest Cretaceous and early Tertiary, although the earliest records are based solely on pollen grains. The first unequivocal appearance of a modern genus is the of *Ulmus* in the early Eocene of western North America. The twigs of *C. lineatum* have petiolate, asymmetrical, lanceolate leaves with acute apices and bases and a prinantky toothed margin. The secondary veins curve towards the margin and end in teeth. The attached fruit is a small winged seed, a *Samara* attached by a short stalk. Other specimens of these leafy twigs are known with small unisexual male and female flowers still attached. The simple structure of the flowers suggest that they were wind pollinated like extant elms. The abundance of *Cedrelospermum* in lake sediments of volcanic areas, together with its production of numerous small winged seeds suggests that it was an early successional coloniser of open spaces. Specimen from The Peabody Museum, Yale University, New Haven (No. 25232). Photo S. R. Manchester, Florida Museum of Natural History, Gainsville, printed by the Photography Department, NMGW.

Plate 119.   The birch family, the Betulaceae, contains six living genera that are usually divided into two tribes: the Betuleae with *Alnus* (alders) and *Betula* (birches) and the Coryleae with *Corylus* (hazels), *Carpinus* (hornbeams), *Ostrya* (hop hornbeams) and the rare genus of two species *Ostryopsis*. This specimen of *Paracarpinus chaneyi* (Lesquereux) Manchester and Crane is from the Oligocene of Oregon in the Unites States (Florida

Museum of Natural History, Gainsville No. 6096; x 2.5). The leaf shows the typical, uniformly spaced straight secondary veins, the toothed margin and intact petiole of extant *Carpinus* and cannot be generically distinguished from the extant species of this genus. They have not been included in *Carpinus* because its nutlets are to be placed in a different genus and named *Asterocarpinus perplexans* Cockerell. The combination of characters of the leaves and associated reproductive organs suggests that this fossil hornbeam-like plant was derived from the same evolutionary radiation as *Carpinus* and *Ostrya* in the late Eocene and early Oligocene. Photo S. R. Manchester, Florida Museum of Natural History, Gainsville, printed by the Photography Department, NMGW.

Plate 120. The Juglandaceae (walnut family) probably had its beginnings in the Upper Cretaceous of Europe where there are bisexual flowers, small nut and pollen suggesting Juglandaceae affinities. The first unequivocal genera of fruits come from the Palaeocene of Europe and North America. One of these is *Polyptera*, an extinct genus with closest affinities to living *Cyclocarya*. The present specimens are part of an enormous collection from the Paleocene of southern Wyoming and eastern Montana. The characteristic fruits, called *Polyptera manningii* Manchester and Dilcher, posses a single wing divided into lobes. In other species the incisions are complete to the nut effectively making several discrete wings. The nut itself is rounded-triangular and tapering from its broad base to a pointed apex. Several of these nuts can been seen here with the consistently associated compound leaves, called *Juglandiphyllites glabra* (Brown ex Walt) Manchester and Dilcher (x 1.5). The 5–7 leaflets are each obovate, with acute apices and toothed margins and are petiolate. Specimen from the Florida Museum of Natural History, Gainsville (No. 13687). Photo S. R. Manchester, Florida Museum of Natural History, Gainsville, printed by the Photography Department, NMGW.

Plate 121. *Florissantia* is a genus of Tertiary fossil flowers whose floral morphology and pollen structure demonstrate that it belongs in the extant Malvales. Based on the fossil record of the distinctive wood and pollen, the evolutionary radiation of the core malvalean families including the Steruliaceae (includes the chocolate trees and camellias), Tiliaceae (limes), Bombaceae (includes the baobabs, silk-cotton trees, kapoks, durians and the balsa wood tree) and the Malvaceae (mallows) arose in the Late Cretaceous or Early Tertiary. The earliest extant genus, *Tilia*, is found as leaves in the Middle Eocene. The present specimen of *Florissantia quilchenensis* (Mathews and Brooke) Manchester comes from the Middle Eocene of Washington State (x 4.5). There are seven relatively large (at least 20 cm in diameter) radially symmetrical flowers borne on long pedicels. The five petals are small and surrounded by the much larger sepals that are fused together into a shallow cup. There are five anthers (two can be seen in the lowest flower) and a central ovary. These flower characters suggest that the flowers were pendant and probably pollinated by insects or birds. The fruits were shed with the persistent sepals so they were likely to have been wind dispersed; a suggestion borne out by their abundance in lake deposits where other kinds of fruits are rare. Specimen from the Stonerose Interpretive Centre, Republic, Washington (No. 87–26–4). Photo S. R. Manchester, Florida Museum of Natural History, Gainsville, printed by the Photography Department, NMGW.

Plate 122. The genus *Populus* (poplars) is widespread today throughout the Northern Hemisphere with 35 species. The earliest recognisable leaves are known from the Palaeocene of North Dakota, USA but is much more widespread in North America in the Eocene.

Populus wilmattae Cockrell comes from the Middle Eocene of northwestern Utah, USA and shows leaves and fruits attached to the same twig (x 0.9). Seven attached leaves are clearly visible and there might be others buried in the sediment. The lower part bears scars showing where leaves have been shed. The leaves are ovate to deltoid with acute to long-attenuate apices, toothed margins and long petioles. The fruiting axis is a raceme, that is with a potentially continuously growing apex with the oldest parts at the base, with ovate three-valved fruits (capsules). The leaf and fruiting racemes confirm the identification of this fossil as *Populus* which belongs to the Salicaceae family that also includes *Salix* (willows) – also been described from the Eocene of North America. The family, therefore, appears to have become essentially modern in most foliage and fruit characters by the Middle Eocene. This contrasts with the abundance of extinct genera known for contemporaneous families of angiosperms. Specimen from the Palaeobotanical Collections of Brigham Young University, Utah (No. 3003). Photo S. R. Manchester, Florida Museum of Natural History, Gainsville, printed by the Photography Department, NMGW.

Plate 123. In Oregon, USA there is an Middle Eocene flora, the Clarno Nut Beds, that has yielded abundant wood, leaves and beautifully fruits and seeds preserved as adpressions, moulds and casts, and permineralisations. Single specimens may be preserved in a combination of modes. The silicification of the seeds and fruits gives a rare opportunity to study the internal structure of these Eocene reproductive organs and enable a much better assessment of their taxonomic affinities. It also provides an important link for comparisons with the relatively abundant European fruit and seed floras at a time when the North American climate was at one of its warmest periods in the Tertiary. Upper left: *Saxifragispermum tetragonalis* Manchester. This is the only species of this extinct genus that is referred to the large family Flacourtiaceae of chiefly tropical shrubs. This fruit is an elongate, four-valved capsule with rows of ovoidal seeds. The specimen illustrated here has part of the fruit wall missing to reveal on locule of the capsule (x 6). Upper right: *Vitis tiffneyi* Manchester. This species belongs to one of the eleven extant genera of the grape family, the Vitaceae and is abundant in the Nut Beds. The illustrated specimen shows the basal view of a seed cast with much of the seed coat flaked away to show the internal mould of the seed cavity. The circular mark shows the position of original attachment (x 16). Middle left: *Magnolia muldoonae* Manchester, preserved as a chalcedony cast (x 11). *Magnolia* leaves are reported from several Tertiary localities in North America but need cuticular characters for verification. There are three species of *Magnolia* from Clarno based on the seeds, *M. muldoonae* being the most common. Middle right: *Eneste oregonense* Manchester and Kress. This seed belongs to one of the three extant genera of the banana family (Musaceae) that today lives in Asia and Africa. These seeds are especially common in Clarno, but they are the only record of the Musaceae from North America. The illustrated seed shows its tapering ellipsoidal shape and its striated coat (x 10). Bottom: *Deviacer wolfei* Manchester. This is a winged seed, a samara, belonging to the Sapindales. These seeds resemble the samaras of living *Acer* (sycamores) although there are several subtle differences that are sufficient to separate them (x 12). They are all from the collections of the Florida Museum of Natural History, Gainsville (Nos 5236, 6533, 6556, 6621, 9859). Photos S. R. Manchester, Florida Museum of Natural History, Gainsville, printed by the Photography Department, NMGW.

Plate 124. These are all small carbonaceous three-dimensionally preserved flowers from the Upper Cretaceous of Sweden and Portugal. The upper two photographs are of

Silvianthecum suecicum Friis (x 50), which can be closely compared to the woody sagifragalean family Escalloniaceae, now growing predominantly in the Southern Hemisphere. The flowers have five small sepals (left upper photo) and five petals (right upper photo). The left middle photograph is of an unnamed flower with an inferior ovary, perianth lobes, pollen sacs and a central style (x 35). To the right of that is a photograph of a single stamen of *Platanus scanius* Friis et al. (x 30). The two photographs on the right of the middle row are of *Esgueiria adenocarpa* Friis et al. (x 60) which is most closely related to the West African genus *Gueira* in the Combreteae. The flowers were borne in dense heads although their arrangement is not known. The illustrated flowers show five sepals, but the petals have been lost. The lower two photographs show a single unnamed flower consisting entirely of stamens. Its systematic affinity is unknown although the structure of its pollen suggests it to be a magnoliid. All the specimens are in the collections of the Swedish Museum of Natural History, Stockholm. Photos E. M. Friis, Swedish Museum of Natural History.

Plate 125. Model showing flowers of a Cretaceous angiosperm of the family Winteraceae. The model was created by Annette Townsend and is now in the collections of the National Museums and Galleries of Wales, Cardiff (V97.26.14). Photo by the Photography Department, NMGW.

Plate 126. A sycamore (*Acer* sp.) from Pleistocene diatomaceous deposits (Pretiglien-Tieglien) near Faufouille, France (x 3). Specimen from the National Museums and Galleries of Wales, Cardiff (No. 86.54G1a). Photo by the Photography Department, NMGW.

Plate 127. The outer coats (nuts) of the hazel, *Corylus avellana* L., from recent postglacial deposits overlying Coed Madog Quarry, Nantlle, Caernarfonshire, Wales (x 3). Specimen from the National Museums and Galleries of Wales, Cardiff (No. 27.110G.1337). Photo by the Photography Department NMGW.

Plate 128. Wood is commonly found in glacial and post-glacial deposits. This cross-section is of *Fraxinus*, which came from post-glacial deposits on the Isle of Wight. Growth rings are clearly visible with the large vessels being in the spring growth. The slide is now in the collections of the National Museums and Galleries of Wales (No. 98.24.G8). Photo. B. A. Thomas, printed by the Photography Department, NMGW.

INDEX

Acer, 118, Pl. 126
Achlamydocarpon, Pl. 35
Achrostichum, 59
Acitheca, 52
Adiantites, Pl. 78
Adpressions, 2–4
Africa, 101
   Northern Africa, 114, 115
   Southern Africa, 47, 75–6, 112–13, 127
Agathis, 83
Alethopteris, 68, 70, Pl. 84
Algae, 1, 3, 12–14, 19
   Dasyclads, 3
Allicospermum, 87
Alloiopteris, Pl. 55
Alnus, 118
Alternating generations, 3, 12
Amber, preservation in, 3
Andrews, Henry, 20, 26, 60, 120, 125
Anemia, 57
Angara Flora, 128
Angiopteris, 54
Angiosperms, 8, 11, 59, 62, 70, 98–107, 115–19, 132–4, 142–6
   Definitions of, 98–9
   Possible ancestors of, 74–5, 93, 95, 99–100, 130
Animal-plant interactions, 101–2
Ankyropteris, 55
Annularia, 46, Pls 41–2
Antarctica, 40, 79, 112, 114, 127–8
Aquatic plants, 74, 102, 103
Araliosoides, 103
Araucaria, 83
Arber, E. A. Newell, 128
Arcellites, Pl. 73

Archaeocalamites, 42, 45, 46
Archaeopteris, 24–6, 79, Pl. 14
Archaeosperma, 66
Arcto-Tertiary forests, 117–18
Arnold, Chester, 129
Arnophyton, Pl. 68
Arthropitys, Pl. 40
Artis, Edmund, 121
Artisia, 79, 121
Asia, 103, 114, 115, 117
Aspidistes, 58
Aspidium, 58
Asplenium, 58
Asterophyllites, 46, Pl. 43
Asterotheca, 52
Asteroxylon, 28, 29, 31, Pl. 17
Atmosphere, composition of, 11, 113
Australia, 28, 40, 75–6, 79, 101, 112, 113, 127–8, 129
Authigenic mineralizations, 2, 4
Autunia, 73–4, Pl. 93
Azolla, 61, Pl. 72
Bacteria, 1, 12
Banks, Harlan, 23, 129
Baragwanathia, 27–8, 129, Pl. 16
Bean, William, 129–30
Beania, 91, 130
   Reconstruction of *Beania*-tree, Pl. 108
Beck, Charles, 24–5, 66
Belgium, 65, 123
Bennettitales, 91–4, 97, 99, 113–16, 130
Bennie, James, 126
Bertrand, Paul, 123
Bevhalstia, 102
Binney, Edward, 124
Biodiversity, past, 9

Botrychium, 54
Botryopteridales, 54–5, 140
Bowerbank, James, 133
Brazil, 101
British Isles, 14, 25, 28–9, 33, 34, 37, 39, 59, 65, 68, 83, 107, 110, 119, 121, 123–4, 126–7, 130, 133, 134–5, 136
Brongniart, Adolphe, 121, 127, 129, 131
Bryophyta, 3, 12, 13–14, 16, 110, 111
Buckland, William, 121, 129
Butterworth, Mavis, 126
Calamites, 43–7, 111, 139, Pl. 39
Calamocarpon, 46
Calamophyton, 40, 51, Pl. 54
Calamopityales, 110
Calamostachys, 46, Pls 45–6
Calder, Mary, 132
Callipteris, 73
Callistophytales, 72–3
Callixylon, Pl. 15
Callospermarion, 73
Canada, 37, 39, 54, 59, 61, 113, 114, 117, 128, 136
Canary Islands, 118
Capitulum, 77
Carboniferous, 6, 7, 10, 11, 21, 27, 32–8, 39, 42–6, 48–9, 52–5, 66–73, 77, 79–81, 83, 86, 89, 101, 110–11, 121–7, 129, 136
Carruthers, William, 130
Casts, 4–6, 36, 46, 52
  Pith casts, 4, 40, 46, 49, 79
Cathaysian Flora, 74, 128
Caulopteris, 52, Pl. 62
Caytonanthus, 95
Caytonia, 95–7, 99, 115–16, 130, Pl. 113
Cedrelospermum, Pl. 118
Chaloneria, 37
Chandler, Marjory, 133
Charcoal, 8, 11, 83, 103, 134
Cheirostrobus, Pl. 49
China, 34, 44, 49, 71, 73–4, 82, 87, 89, 91, 110–13, 144

Chinle Formation Flora, 131
Chlamydomonas, 3
Cladophlebis, Pl. 67
Cladoxylales, 40, 41, 42, 51, 140
Clarno Nut Bed, 107
Classopollis, 86
Climate, 7, 11, 75
Club mosses, *see* Lycophyta
Coal and peat, 6–7, 11, 33–5, 52, 81, 111–14, 118, 122–7
Coal balls, 6, 7, 35, 124–5
Coenopteridales, 51, 54, 140
Collinson, Margaret, 133, 136
Compressions, 2–5
Conifers, *see* Pinales
Coniopteris, 58
Conservation of sites, 134–5
Cookson, Isabel, 129
Cooksonia, 16–20, 28, 129, Pl. 3
Cordaitanthales, 73, 78–81, 83, 85, 89, 111–12, 127
Cordaitanthus, Pl. 99
Cordaites, Pl. 99
Corylus, Pl. 127
Corystospermales, 95–7
Crane, Peter, 104, 134
Crepet, William, 131
Cretaceous, 8, 50, 57–62, 86, 91–7, 100–5, 115, 130, 132–4
Cretaceous-Tertiary extinction, 62, 86, 91, 93, 97, 105, 116–17, 134
Crookall, Robert, 123
Cross, Aureal, 125
Crossothecales, 54, 140
Crossozamia, Pl. 106
Crow, Francis, 133
Croziers, of fern fronds, Pl. 61
Culmitschia, Pls 100–1
Cupressinocladus, 87
Cupule, 63, 65–77, 95, 125
Cuticles, 3, 5, 9, 13–4, 27, 31, 33, 70, 74, 86, 91, 93, 114, 131, 137

Cyathocarpus, 52, 54, Pl. 59
Cycadales, 62, 63, 71, 89–91, 93, 94, 99, 110–15, 121, 130
Cycadeoidea, 130, Pl. 112
Cycadophytes, 112
Cycas, 89, 91
Cyclostigma, 34, Pl. 20
Czech Republic, 18
Darrah, William, 123, 125
Darwin, Charles, 122
Davallia, 58
Davies, David, 123
Dawson, John, 128–9
Dawsonites, 23
Deciduous, 7, 105, 115–18, 134
Delevoryas, Theodore, 90, 131
Dennstaedtiopsis, 59
Desert conditions, 114, 115
Deviacer, Pl. 123
Devonian, 3, 6, 8, 13–31, 34, 40–1, 49–51, 62, 64–5, 109–10, 128–9
Diaphragm, stem, 40
Dicksonia, 58
Dicksonites, 72
Dicroidium, 97, Pl. 114
Dijkstra, Sijben, 126
Dinosaurs, 101, 116
Diplopteridium, Pl. 79
Dix, Emily, 123
Dwarf shoots, 83
Edwards, Dianne, 18
El Chichon, volcanic eruption, 116
Elaters, 40, 42, 46, 49
Elatides, Pl. 104
Elkinsia, 64–5
Embryos, 3, 12, 63, 98
Eneste, Pl. 123
Ephedra, 99
Equisetales, 41–9, 139
Equisetum, 40, 42, 46–7, Pl. 53
Esgueiria, Pl. 124
Eskdalia, 33, Pl. 19
Estinnophyton, 30–1

Euramerian Flora, 128
Eusphenopteris, Pl. 82
Ferns, *see* Pteridophyta
Fertilization, double, 98
Filicales, 51, 55–9, 114, 140
Flemingites, 35, Pl. 34
Florin, Rudolph, 132
Florissantia, 106, Pl. 121
Flowering plants, *see* Angiosperms
Forests, fossil, 7, 37, 124, Pls 26–7
Fossil record, bias in, 8
France, 121, 123
Fraxinus, Pl. 128
Friis, Else Marie, 103, 134
Galium, 49
Gametophytes, 3, 12, 14, 17, 26, 62–5
Gangamopteris, 75, 128
Gaspé, 128–9
GEOSITES, IUGS, 135
Germany, 39, 113, 121, 123
Gigantonoclea, Pls 95–6
Gigantonomiales, 74–5, 111–12, 141
Gillespie, William, 64, 136
Ginkgo, 77, 82, 87–9
Ginkgoales, 62, 77, 82, 97–9, 113–15
Ginkgoites, 89, Pl. 105
Girdling leaf traces, 89, 93
Glaciers, 118, 130
Glands, epidermal, 5
Glossopteridales, 75–9, 95, 99, 111–12, 141
Gnetales, 93, 99, 100, 142
Gondwana, 46, 75, 83, 127–8
Göppert, Heinrich, 122, 127
Gordon, William, 125
Gosslingia, 21, Pls 9–10
Gothan, Walther, 123
Grass, 106
Greenland, 130
Gymnospermophyta, 36, 62–97, 99–100, 109–17, 127, 132, 141
Hairs, epidermal, 5, 33, 46
Halle, Thore, 130
Halonia, Pl. 23

Harris, Tom, 91–2, 96, 130
Hausmania, 99
Helminthostachys, 54
Heterospory, 26, 31, 35, 37, 42, 46, 50–1, 61, 63, 109
Heuber, Fran, 129
Histiopteris, 59
Holttum, Richard, 130
Homospory, 26–7, 29, 31, 46, 50, 54, 109
Hooker, Joseph, 124
Horneophyton, 16–17
Horsetails, *see* Sphenophyta
Hoskins, J. Hobart, 125
Hughes, Norman, 132
Huperzia, 27, 29, 31
Hutton, William, 121, 124, 129
Hydathodes, 70
Hydrasperma, Pls 75–6
Hypodermis, 31
Ibrahim, A. C., 126
Ice ages, 77, 87, 100, 106, 110, 112, 118
Impressions, 3, 9
India, 75–6, 85, 112, 114–16, 127
Indusium, 58
Integument, 36, 63, 67, 73, 98
International Organisation of Palaeobotany, 136
Isoetes, 37
Janssen, Raymond, 136
Japan, 34, 59, 125
Jennings, James, 136
Jongmans, Wilhelmus, 123–4, 126, 136
Jurassic, 6, 33, 54, 55, 57–8, 85–7, 93, 101, 114, 124, 127, 129–32, 136
Kazakhstan, 18, 112
Kidston, Robert, 15–16, 28–9, 123, 126, 129–30
Kidstonophyton, 17
Klukia, 57
Kosanke, R. M., 126
Krakatau, volcanic eruption, 116
Kremp, Gerhard, 126
K/T, *see* Cretaceous-Tertiary extinction
Lacey, William, 17, 125

Lagenicula, 35, Pl. 34
Lagenoisporites, 35
Lagenostomales, 64–8, 110–11, 141
Lagenostome, 63, 65, 68, 71, 73
Lanceolatus, 77, Pl. 97
Land, adaption to life on, 13–14
Lang, William, 15–18, 28–9, 125, 129
Laveineopteris, 70
Leaves
    Leaf cushions, 33–5, Pls 21, 24–5
    Leaf scars, 34, 52, Pls 21, 62
    Origins of, 21
Lepidocarpon, 35, Pls 35–6
Lepidodendron, 9, 10, 33–5, Pls 21–2, 29
Lepidophloios, 35, Pls 18, 24–5
Lepidostrobophyllum, Pl. 35
Lepidostrobus, Pls 32–4
Leptocycas, 90
Leptostrobales, 113–15, 141
Lesqueria, 104
Lianas, 67–70, 74
Libya, 18
Lidgettonia, Pl. 98
Ligule, 31–5, 37
Liliopsida, 100
Lindley, John, 121, 124, 129
Linopteris, 71, Pl. 90
Liquidamber, 118
Liriophyllum, 105
Lister, Martin, 120
Llwyd, Edward, 120
Lobatopteris, 52, 54, Pl. 60
London Clay Flora, 6, 59, 107, 117, 132–3, 136
Long, Albert, 125
Lumen, cell, 5
Lyonophyton, Pl. 2
Lycophyta, 3, 7, 9, 12, 21, 27–39, 49, 109–12, 138–9
Lycopodium, 27, 31, 37
Lycospora, Pl. 34
Lygodium, 57, Pl. 70
Macroneuropteris, 70
Macrostachya, 46, Pl. 47

Macrozamia, 89, 93
Magdefrau, Karl, 37
Magnolia, 104, Pl. 123
Magnoliophyllum, 105
Mamay, Sergius, 124
Marattia, 54
Marattiales, 51, 52, 54, 111, 140
Mariopteris, 67–8, Pls 80–1
Marsilea, 61
Martin, William, 120
Massulae, 61
Mazon Creek Flora, 136
McCandlish, Keith, 136
Medullosales, 68–73, 89, 141
Megaspores, 26–7, 35, 37, 61, 126
Meristem, 25
Metaclepsydropsis, Pls 56–7
Metasequoia, 59, 82, 115
Mexico, 116, 134
Meyen, Sergei, 128
Microphylls, 27, 31, 50
Micropyle, 63, 67, 71, 83
Microspores, 26–7, 35, 37, 42, 46, 61, 63, 126
Molecular clock, 8
Monanthesia, 94
Mongolia, 112
Monocotyledons, *see* Liliopsida
Morgans, Helen, 136
Moulds, 4
Naming plant fossils, 9
Nathorstiana, 37
Nature Reserves, 135
Neolithic, 10
Neuralethopteris, Pl. 83
Neuropteris, 69–70, Pls 87–8
New Zealand, 40
Nicol, William, 124
Nilsonia, 92, Pl. 107
Nilssoniopteris, Pl. 111
Nipa, 118
Noé, Adolphe, 125
Nomenclature, 9–10, 126

North America, 6, 21, 25, 44, 52, 55, 59, 61, 64–5, 71, 74, 82, 83, 87, 95, 102–3, 107, 110–18, 122, 124, 127, 131, 133, 134, 136
Nothia, 17
Nothofagus, 115
Nuclei, 1, 98
Odontopteris, 70
Oil resources, 11
Onoclea, 59
Ophioglossales, 51, 54, 140
Ophioglossum, 54
Osmundales, 55, 57, 140
Osmundopsis, Pl. 67
Otozamites, Pl. 110
Ovules, *see* Seeds and ovules
Oxroadia, Pl. 76
Pachypteris, 33, 97, 101, Pl. 115
Palaeobotany
    Beginnings of, 120–2
    Future for, 135–7
    History of, 120–37
    Role of, 10–11
Palaeofloristics, 10, 108–19, 127–8
Palaeosmunda, Pl. 66
Palaeostachya, 46, Pl. 44
Paleocarpinus, 59
Palms, 115, 146, Pl. 116
Paper coals, 33, Pl. 19
Paracarpinus, Pl. 119
Paralycopodites, Pl. 18
Pararaucaria, 86
Parichnos, 34, Pl. 21
Paripteris, 71, Pls 89, 91
Parka, 19, Pl. 5
Parkinson, James, 120
Parson, James, 132
Patagonia, 86
Paurodendron, 39
Peat, *see* Coal and peat
Pecopteris, 55, Pl. 64
Pedersen, Kaj, 134
Peels, acetate, 5, 125

Peltaspermales, 73–4, 82, 89, 95, 97, 111–15
Permian, 21, 42, 44, 46–7, 49, 50, 55, 62, 71, 73–7, 81, 83, 85, 89, 91, 95, 105, 111–13, 123, 127
Permian-Triassic extinction, 62, 74, 85, 112
Petrifactions, 2, 5–6, 9, 15, 28–9, 114, 124
Petrified trees, 131, 134
Petry, Loren, 129
Phlebopteris, 57, Pl. 71
Phloem, 13, 25, 98
Photosynthesis, 1, 46
Phyllotheca, 47
Phytoleim, 3
Pilularia, 61
Pinales, 6, 8, 25, 59, 62, 63, 67, 73, 75, 78–9, 82–7, 89, 99, 101, 105, 110–17, 127, 131
Pinus, 82
Pitus, 66, 125
Plants, definition of, 1–3
Platanus, 102, 103, Pl. 124
Pleistocene, 87, 118
Pleuromeia, 37–8, 113
Pollen, 5, 8, 12, 63, 73, 77, 79, 83, 86–7, 95, 98–106, 116–19, 122, 127, 132–4, 137
  Tectate pollen, 98–9
Pollen-sacs, 79
Pollination
  By insects, 101–2, 105
  By wind, 102
Polymorphopteris, 52
Polypodium, 58
Polyptera, Pl. 120
Populus, 118, Pl. 122
Portugal, 103, 134
Pothocites, 42, Pl. 44
Potonié, Robert, 126–7
Potoniea, 71, Pl. 92
Pre-pollen, 63, 65, 67, 71
Preservation, modes of, 2–9, 33, 110
Progymnospermophyta, 23–6, 50, 79, 109–10, 141
Protea, 115
Protoctists, 1, 3, 12

Protocalamostachys, Pl. 76
Protolepidodendrales, 31, 139
Protopteridium, 25, Pl. 12
Psaronius, 52–3, Pl. 63
Pseudobornia, 41–2, Pl. 38
Pseudosporochnus, 51
Psilophyton, 21, 23, Pl. 11
P/T, *see* Permian-Triassic extinction
Pteridophyta, 3, 12, 17, 21, 23, 40–1, 50–62, 99, 109–16, 123, 130, 140
Pteridosperms, 65–74, 110–14, 123, 125
Pteris, 58
Pteronilssonia, 75
Pterophyllum, Pl. 109
Pullaritheca, Pl. 76
Quaternary, 106, 118
Rain forests, 37, 70, 105, 111, 118
Raistrick, Arthur, 126
Rajmahal Hills, India, 114
Raniganjia, 46, Pl. 48
Ray, John, 120
Read, Charles, 124
Regnellidium, 61
Reid, Mary, 133
Reinsch, P. F., 126
Remy, Winfred, 17
Renaultia, 54
Renier, Armand, 123
Reticulopteris, 70
Rhacophytales, 51
Rhynia, 14, 16–17, Pl. 1
Rhynie Chert, 6, 14–17, 28–31, 129, Pls 1–2, 17
Rhyniophyta, 14–19, 21, 27, 109, 138
Rhyniophytoid, 18–19, 109
Riparian vegetation, 7–8
Roots, 27, 36, 52, 59–60, 132
Russia, 33, 103
Sageopteris, 95, 99, 130
Salvinia, 61
Sapindopsis, 105
Sassafras, 103
Sawdonia, 21–2
Saxifragispermum, Pl. 123

Index

Schimper, Wilhelm, 122
Schlotheim, Ernst von, 121, 133
Schopf, James, 125–6
Scolecopteris, 52
Scott, Dunkfield H., 78, 125, 128, 133
Secondary growth, 13, 25, 36, 47, 52
Seeds and ovules, 3, 36, 62–107, 109, 111, 116–17, 121, 125, 130–3
Selaginella, 31, 37, 39
Selaginellites, Pl. 37
Sequoia, 82, 115
Seward, Albert, 128, 130
Siberia, 81, 112–15, 118, 128
Sigillaria, 35, Pl. 31
Sigillariostrobus, Pl. 32
Silurian, 3, 8, 12, 14, 16–19, 28, 109
Silvianthecum, 103
Smith, Harold, 126–7
South America, 18, 75–6, 85–6, 101, 110, 112, 114–15, 127, 131, 144
Sphenophyllum, 48–9, 111, Pls 50–2
Sphenophyta, 3, 4, 12, 40–50, 79, 109–15, 139
Sphenopteridium, Pls 74, 77
Sporangia, 4, 16, 18–21, 23, 25–31, 34–5, 37, 40–2, 46–7, 50–2, 55, 57–9, 61, 63, 71, 73, 97
    Annulus, 52, 55, 57–8
    Dehiscence, 16, 20–1, 25, 29, 52, 57
Sporangiophores, 40–2, 46–7, 49, 73
Spores, 4, 5, 7–8, 11–12, 14, 16, 18–21, 25–7, 29, 31, 35–6, 40, 42, 49, 51–2, 54–5, 58, 116, 118–19, 122, 126–7, 131, 137
Sporophylls, 27, 34–5, 37, 83, 91, 93
Sporophytes, 3, 12, 14, 17
Sporopollenin, 5, 14, 19, 61
Stauropteris, Pl. 58
Steganotheca, 19, Pl. 4
Stele, 13, 16, 21, 49, 51, 65–6, 68, 89
Sternberg, Kaspar von, 36, 121, 131, 133
Sterome, 13
Stewart, Wilson, 125
Stigmaria, 9, 36, 37, 124, Pls 25–9, Pl. 40

Stockey, Ruth, 132
Stomata, 5, 11, 14, 16, 18, 20, 33–4, 46, 70, 83, 91, 114
Stopes, Marie, 125, 127
Stratigraphy, 108, 123–4, 127
Supaia, 74, Pl. 94
Svalbardia, 25–6, Pl. 13
Swamp vegetation, 11, 33, 35, 39, 52, 54, 79
Synangium, 67, 71
Taphonomy, 6–7
Tempskya, 60
Tertiary, 6, 47, 50, 54, 57–62, 82, 87–91, 101–2, 105–7, 116–18, 121, 132–4, 136
Tethys Sea, 117
Tetracentron, 99
Tetrastichia, Pl. 76
Thomas, H. Hamshaw, 130–1
Tidwell, Don, 136
Todites, 57
Tree ferns, 51–3, 58–60, 68, 91, 111
Trees, 7, 13, 25, 33, 37, 43, 52, 67–8, 74–80, 82, 109, 111, 115, 117, 131
Triassic, 8, 37, 38, 57–8, 62, 74, 85–97, 99, 112–13, 122, 131
Trigonocarpales, *see* Medullosales
Trigonocarpus, Pls 85–6
Trimerophytopsida, 21, 23, 25, 26, 40, 138
Tropical vegetation, 27, 32, 37, 43–9, 53–8, 65–74, 77–81, 105, 111–18, 133
Ukraine, 125
Ulmus, 118
Upchurch, Gary, 134
Urnatopteridales, 54, 140
USA, 57, 61, 65, 83, 102–3, 112–13, 117, 123, 125–6, 131, 133–6
Variscan Orogeny, 111
Vascular plants
    Classification of, 138–46
    First, 17–19, 128–9
Vitis, Pl. 123
Voltzialeans, 85
Wagner, Robert, 120, 124
Wales, 18–19, 21, 123

Walton, John, 124–5
Wegener, Alfred, 10, 75, 122, 128
Weichselia, 56–7, 114, Pl. 69
Weiland, George, 131–2
Welsh Borderlands, 18
Welwitschia, 99
Williamson, William, 36, 124, 129–30
Wolfe, Jack, 134
Woodward, John, 120
Xylem, primary, 13, 25, 31, 36, 51, 72
Xylem, secondary, 4, 10, 25, 33, 42, 45, 65–8, 72, 75, 79, 83, 93, 98–100, 109, 118, 124, 131, Pls 102–3
Zamia, 89
Zechstein Sea, 111
Zeiller, René, 123
Zeilleria, 54, Pl. 65
Zodrow, Erwin, 120, 136
Zosterophyllopsida, 20–2, 138
Zosterophyllum, 20–1, Pls 6–8

PLATES

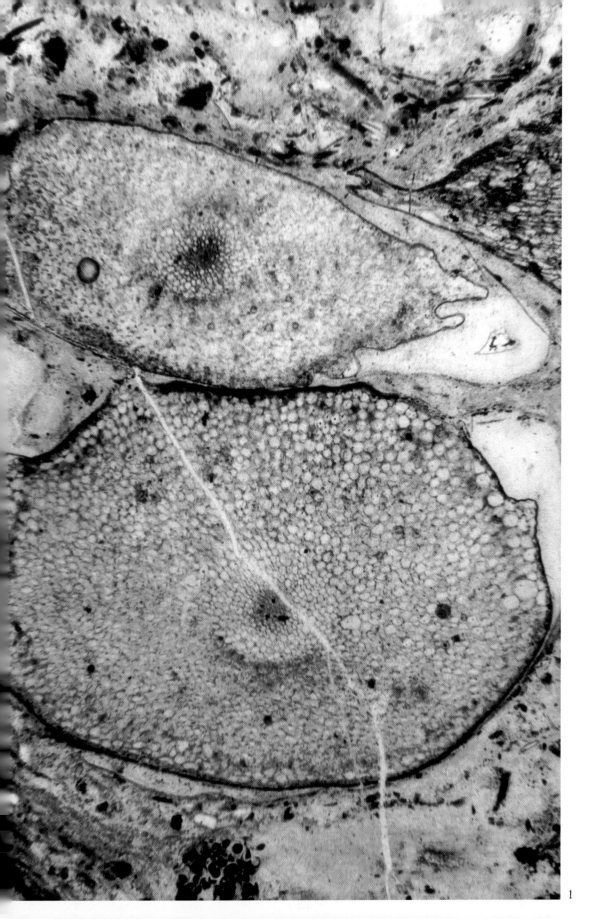

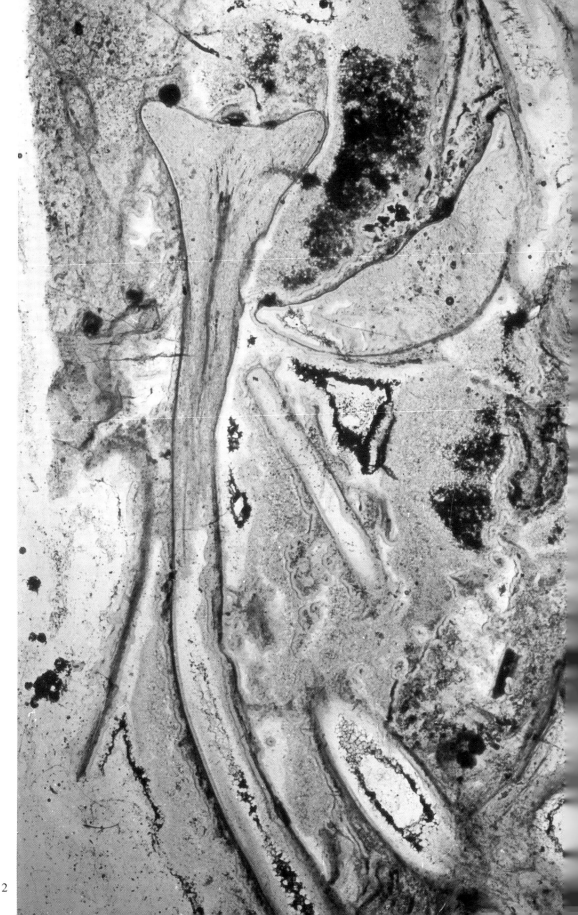

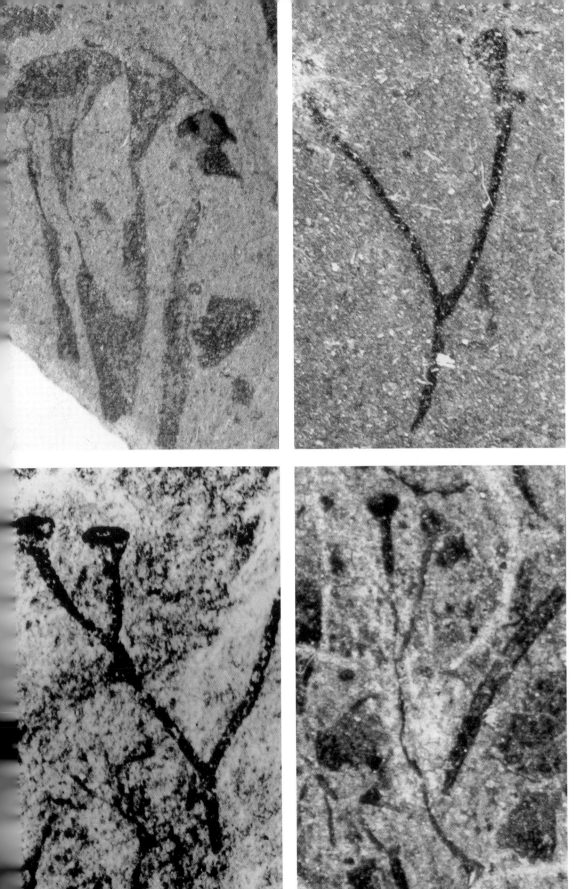

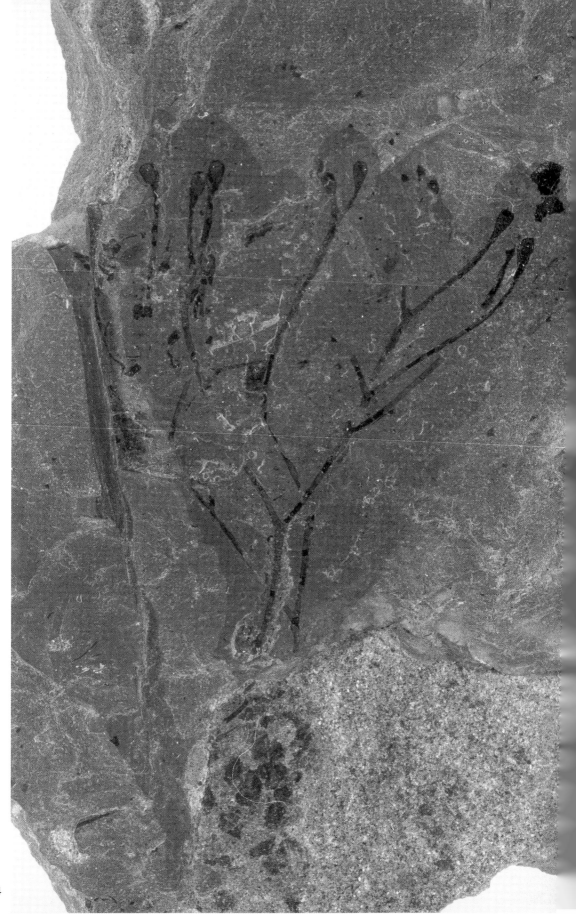

4

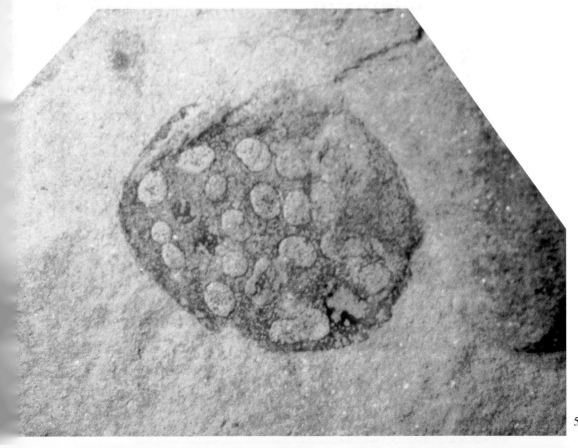

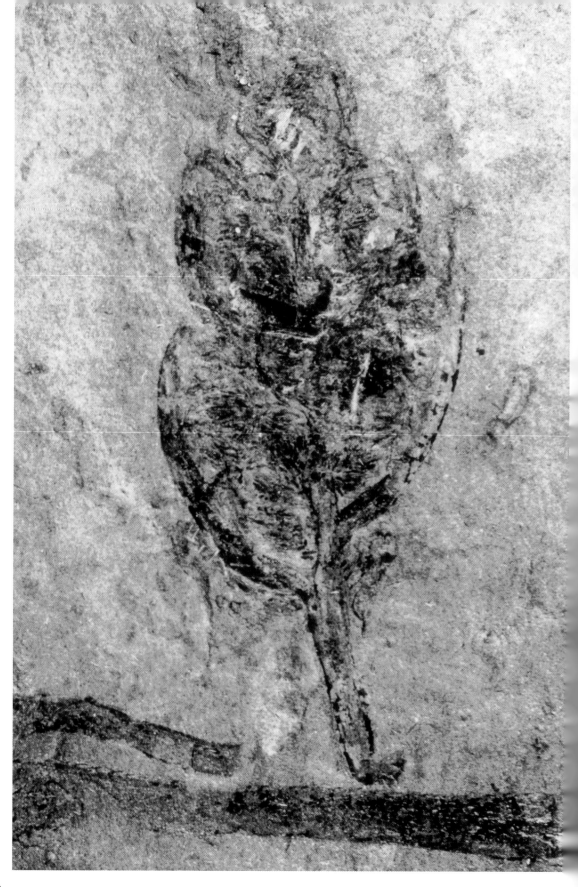

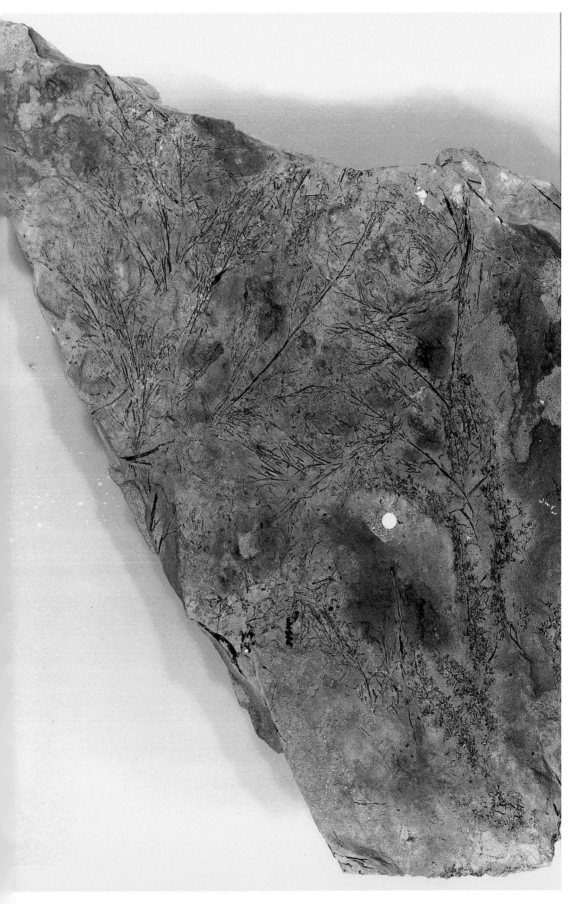

14

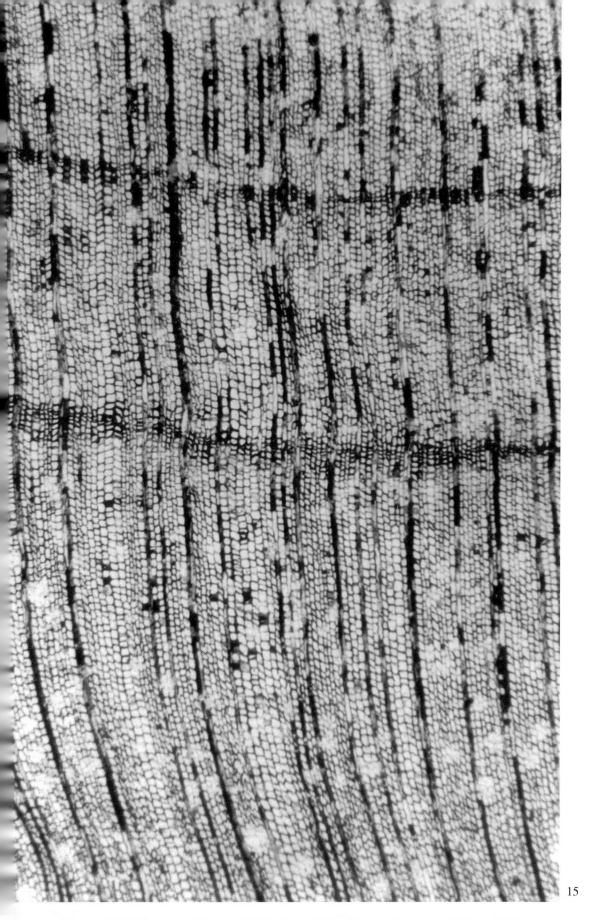

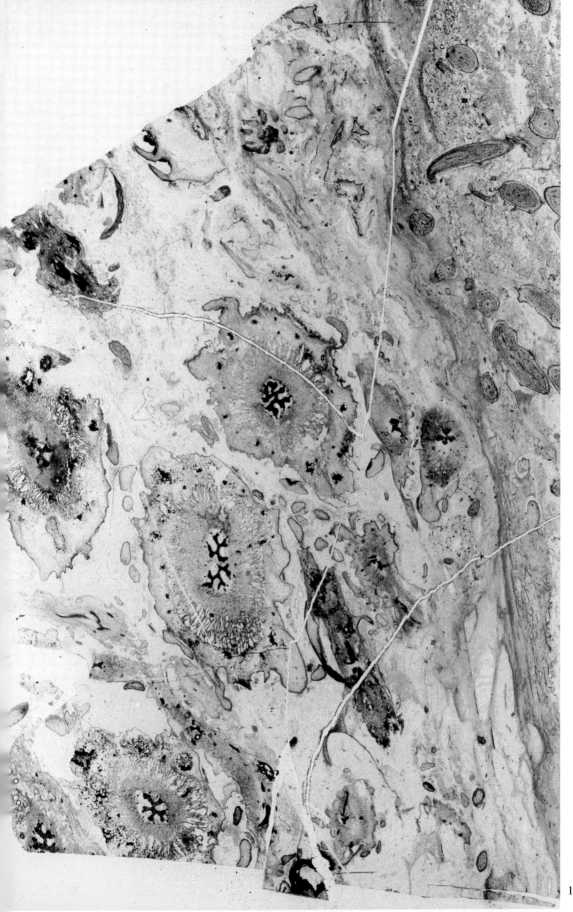

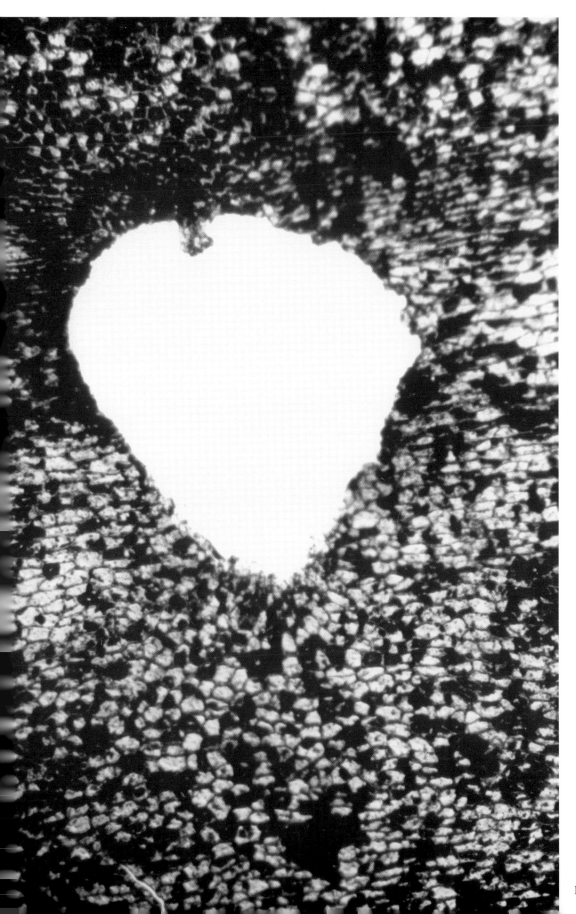

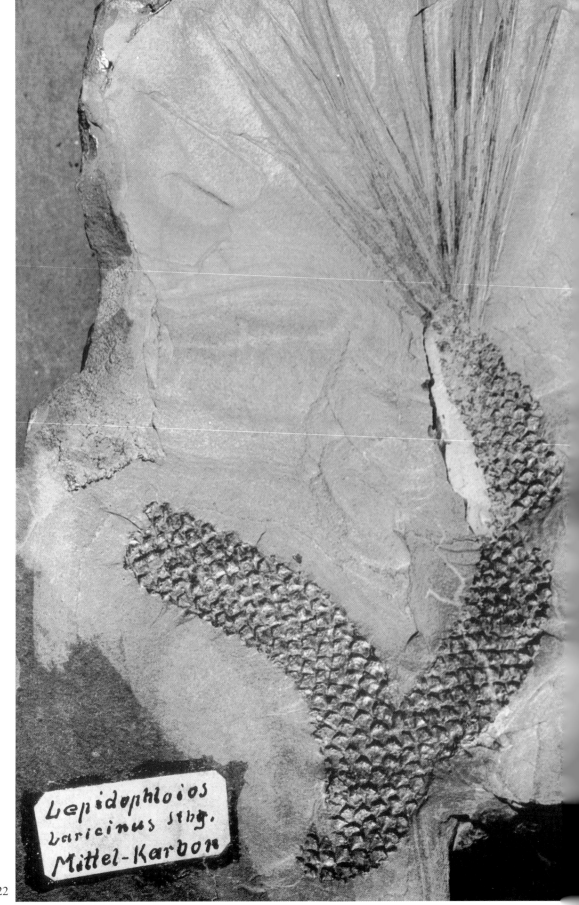

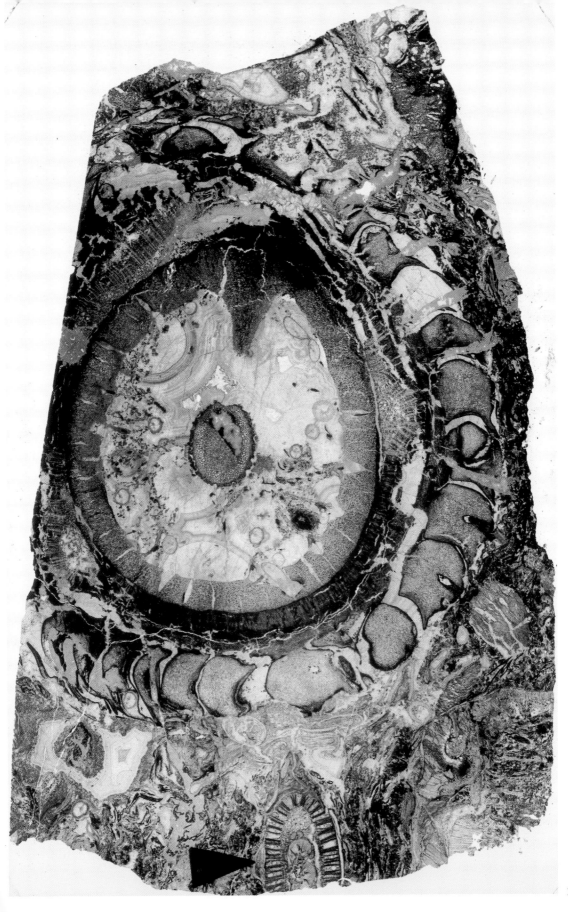

25

27

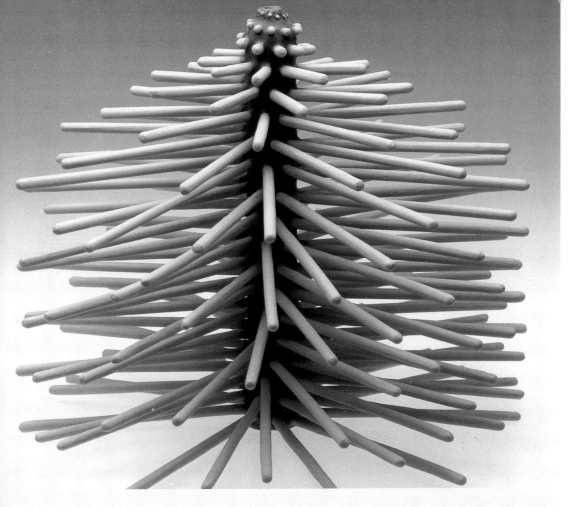

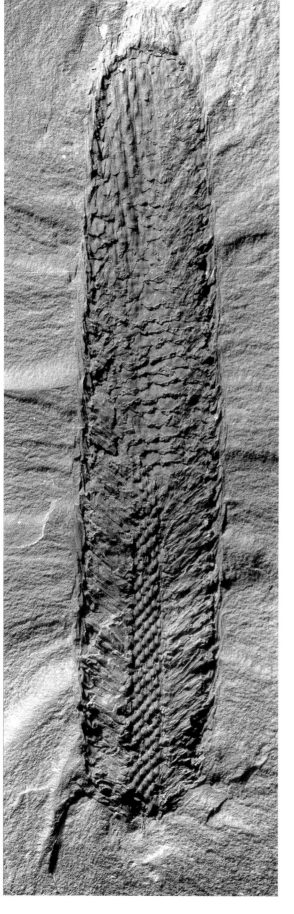
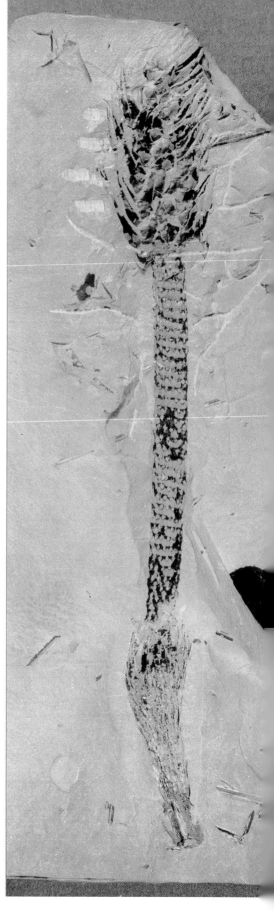

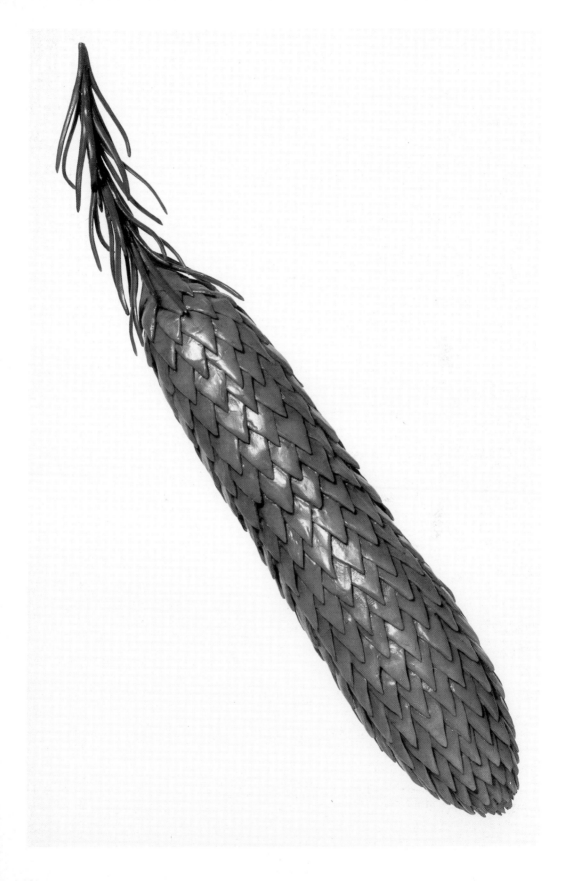

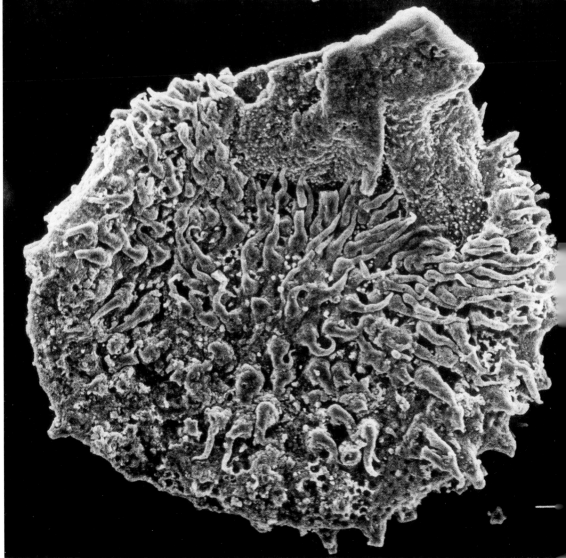

35

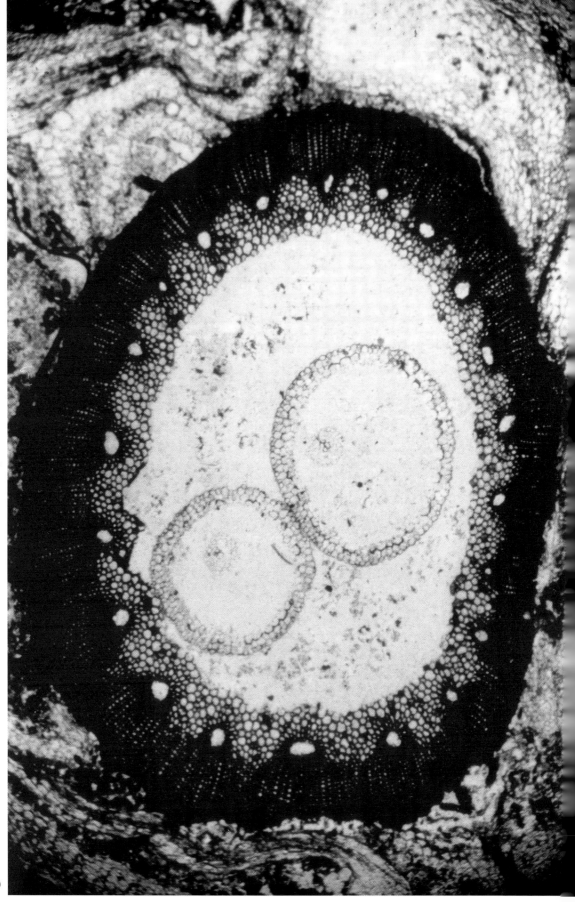

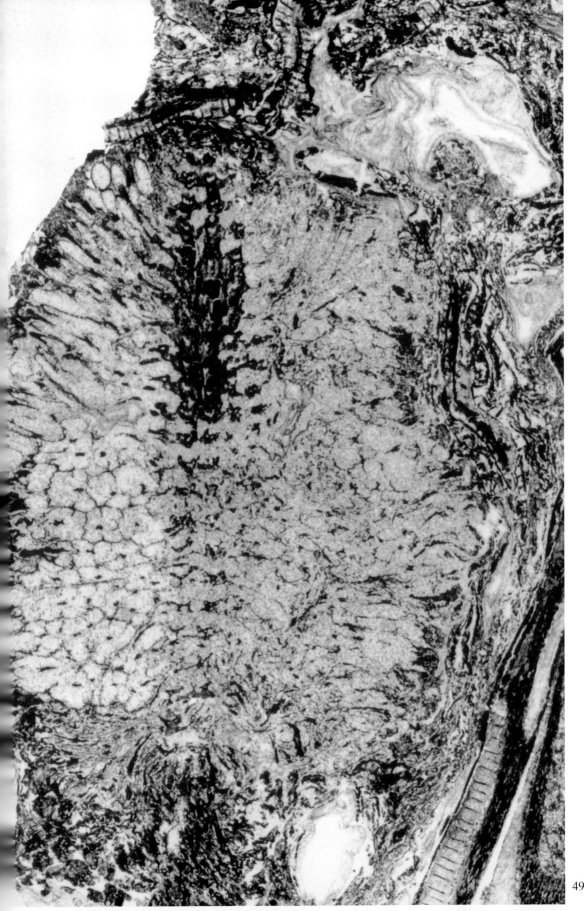

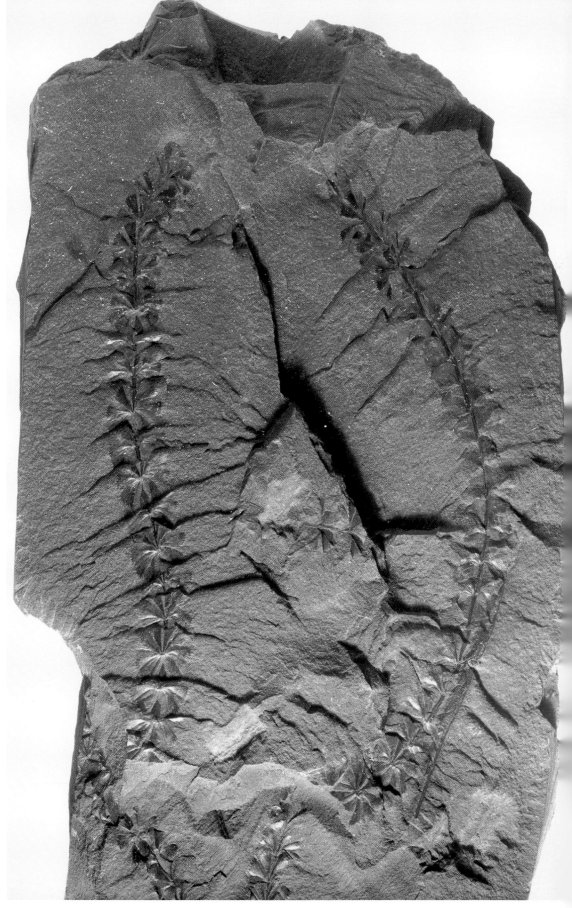

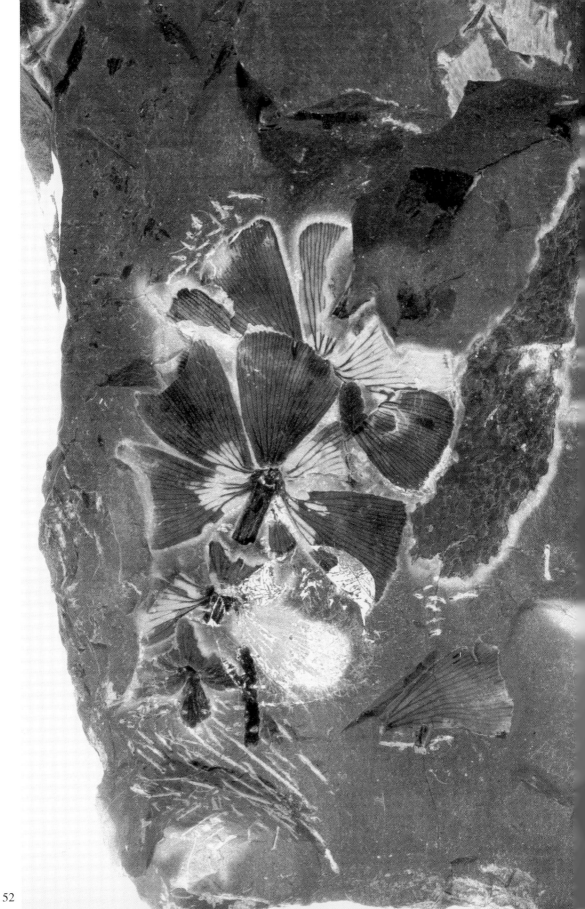

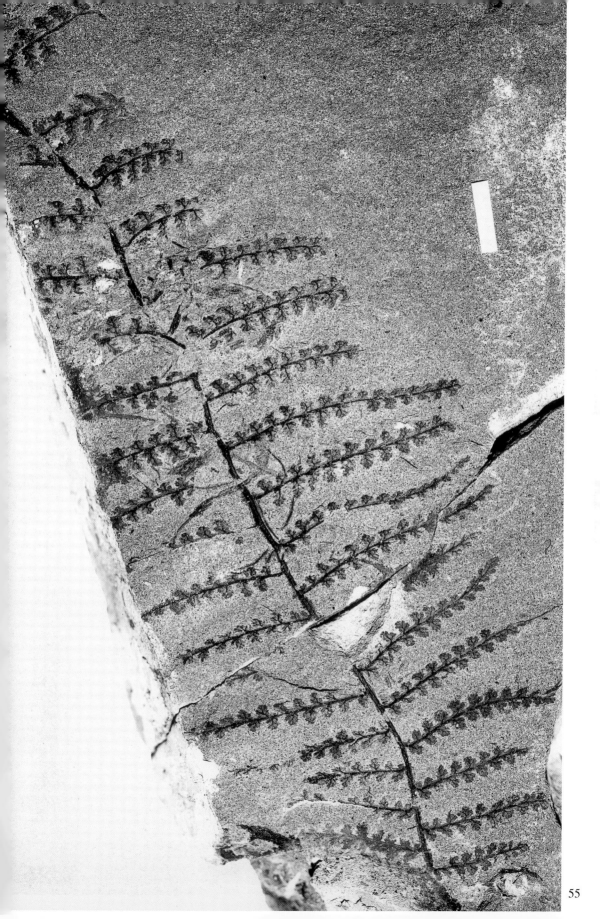

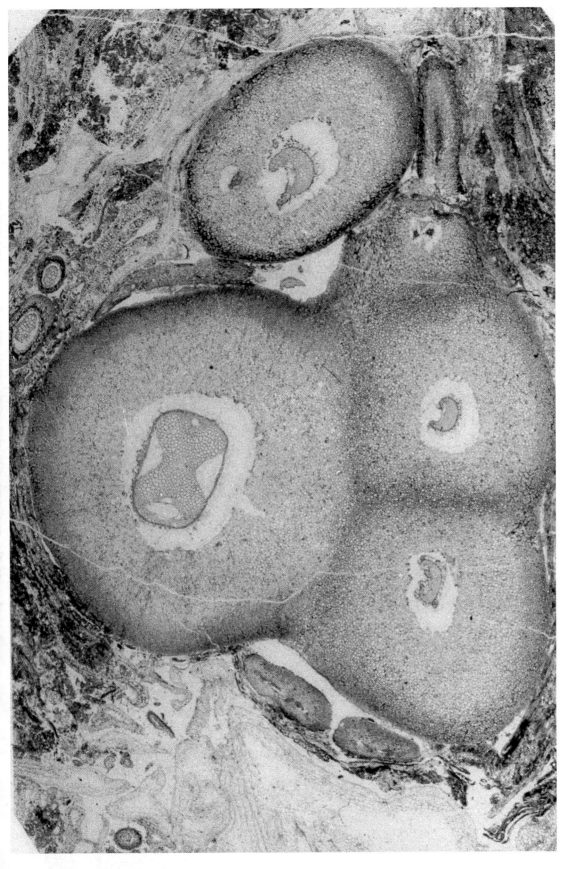

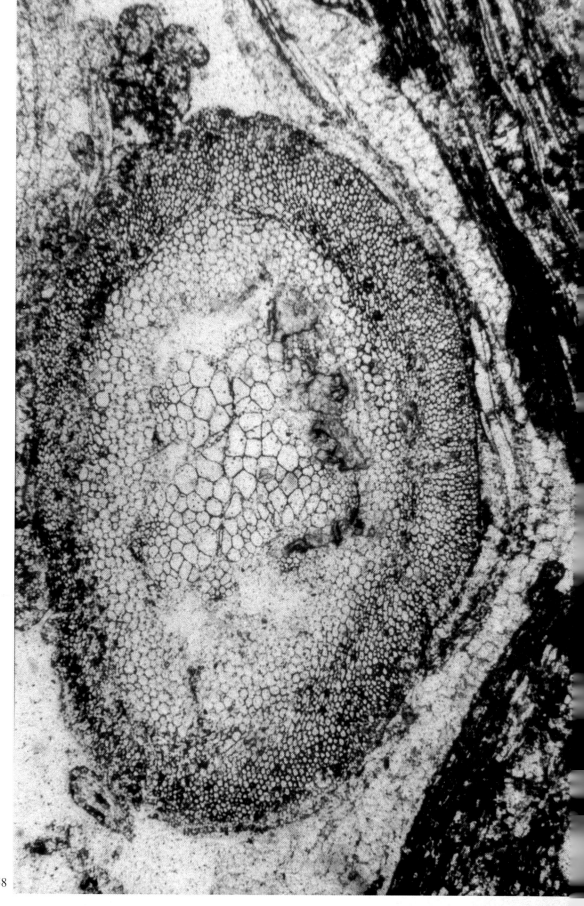

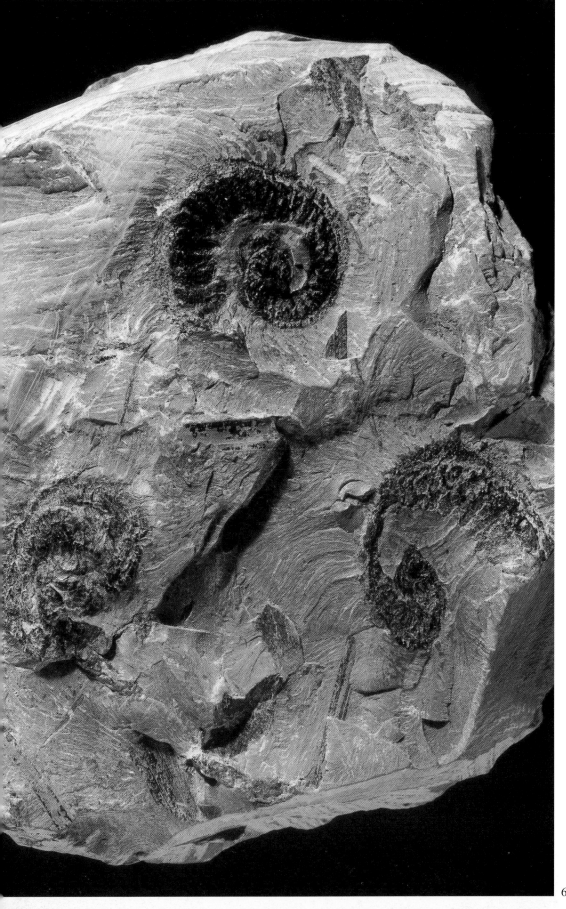

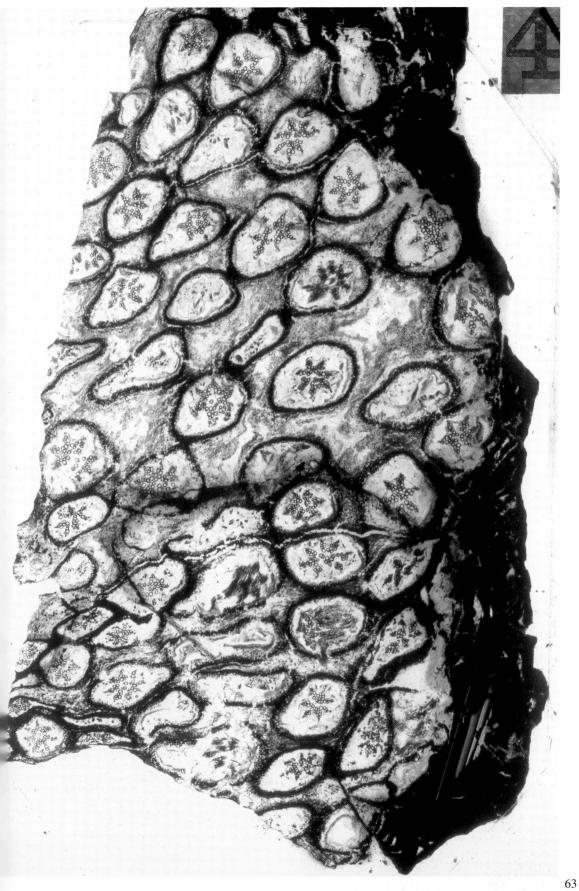

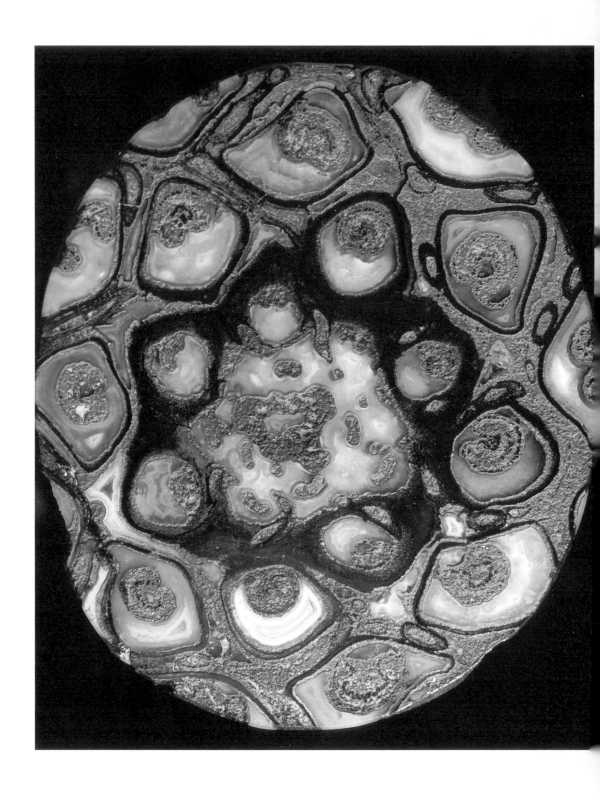

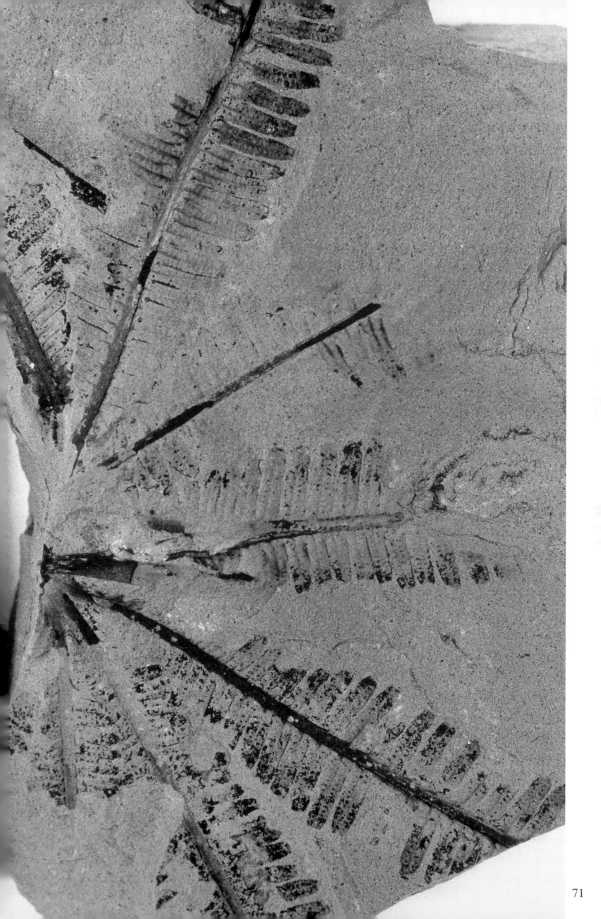

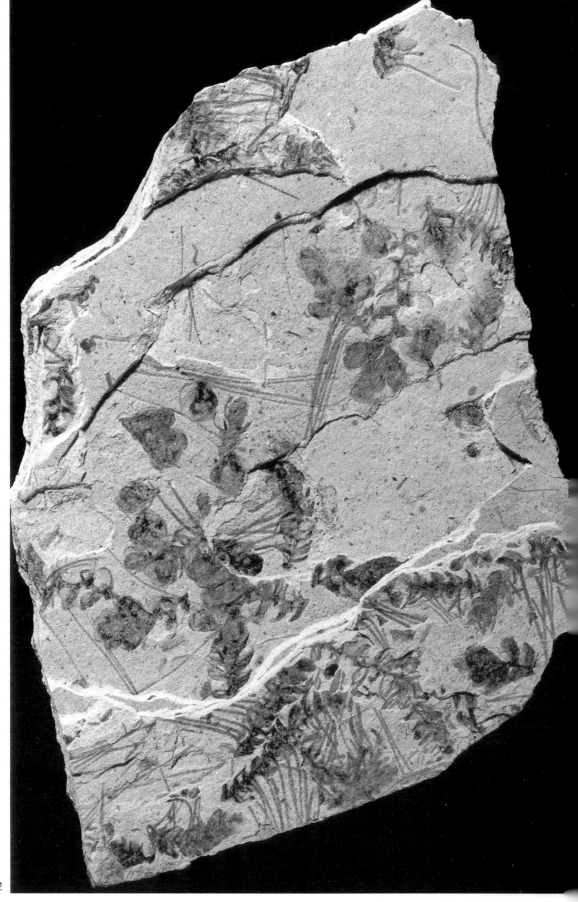

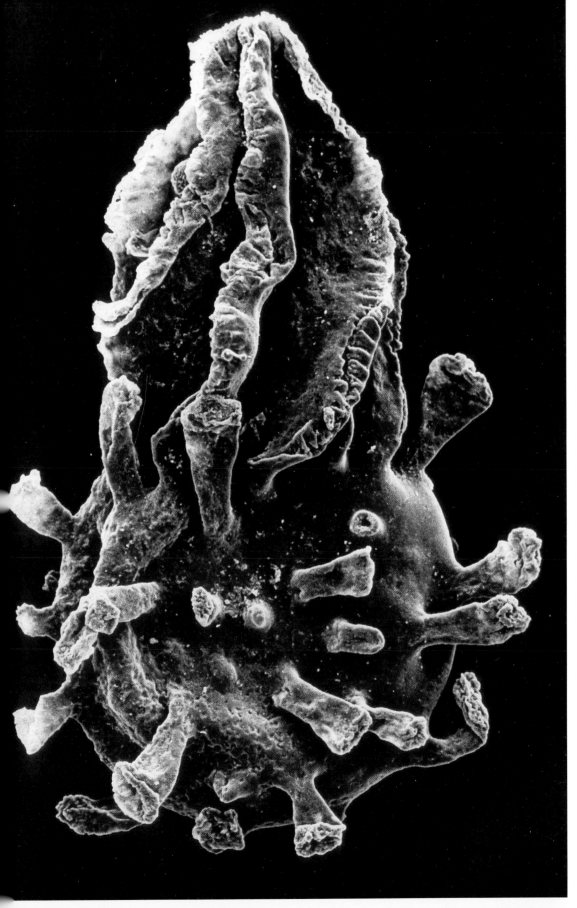

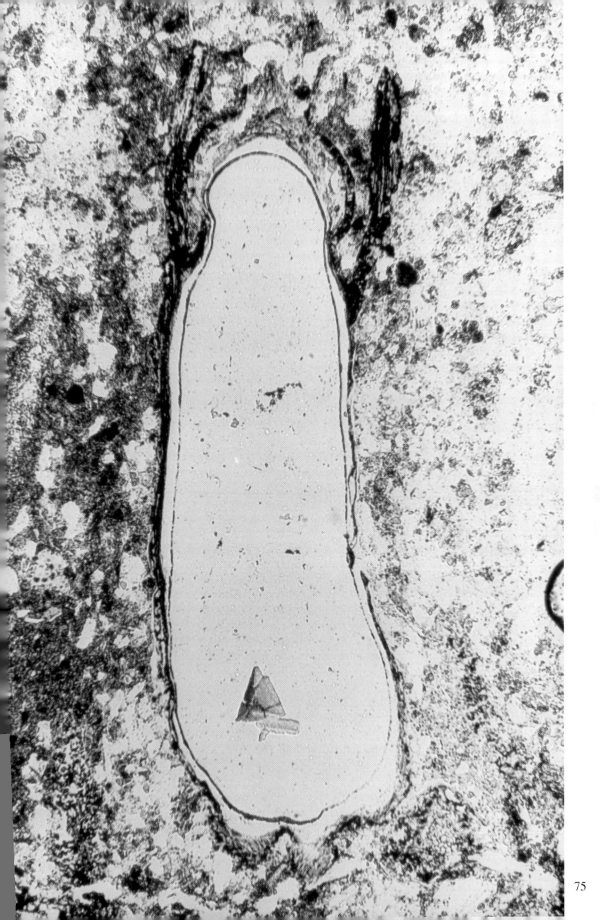

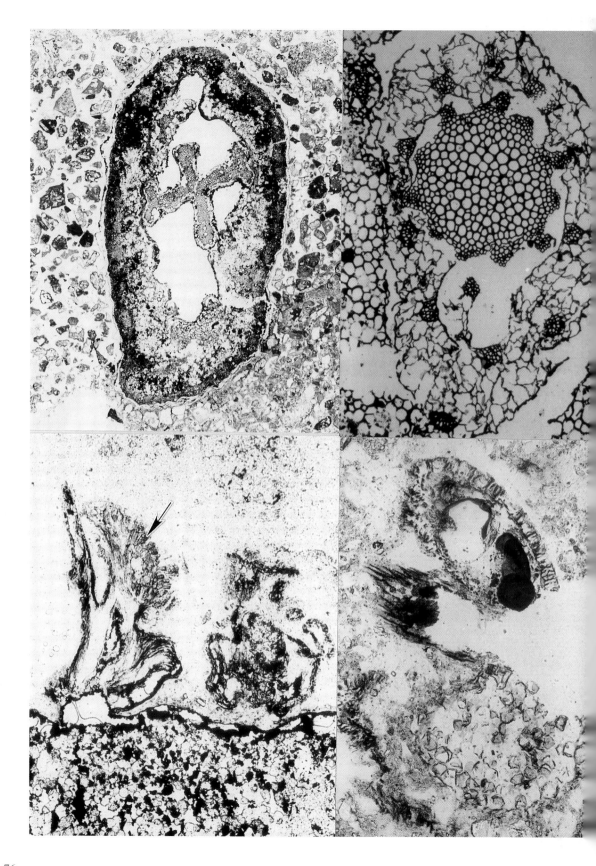

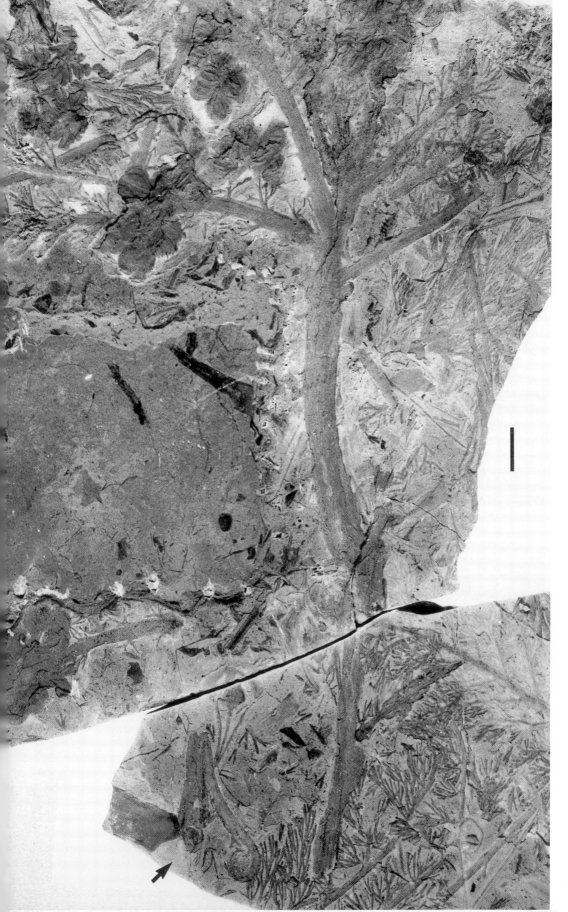

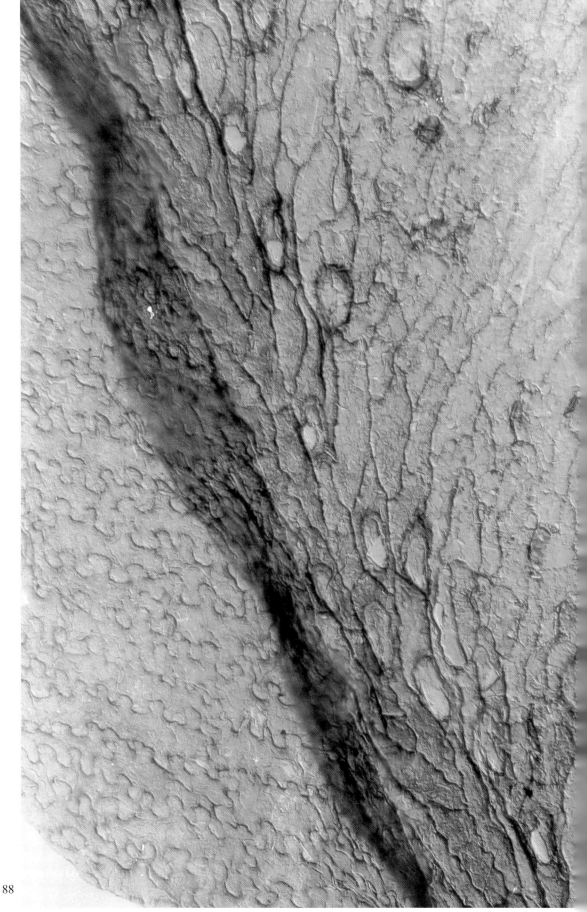

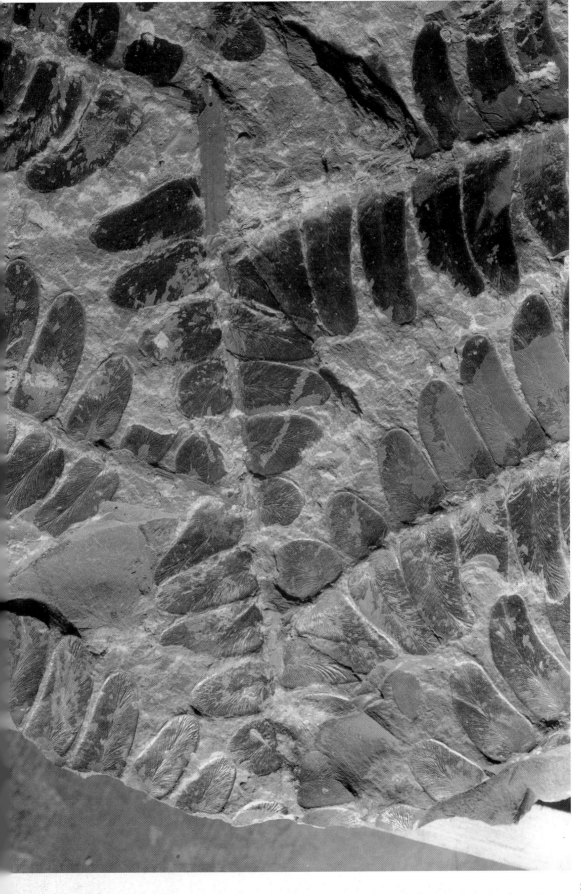

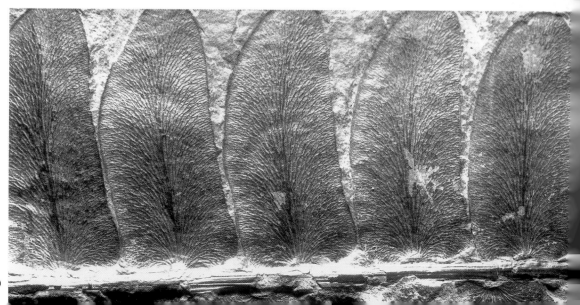

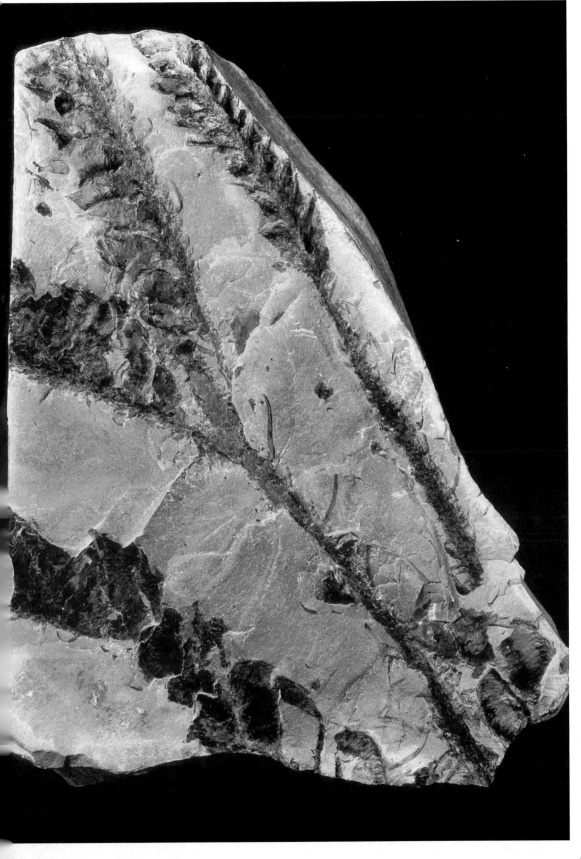

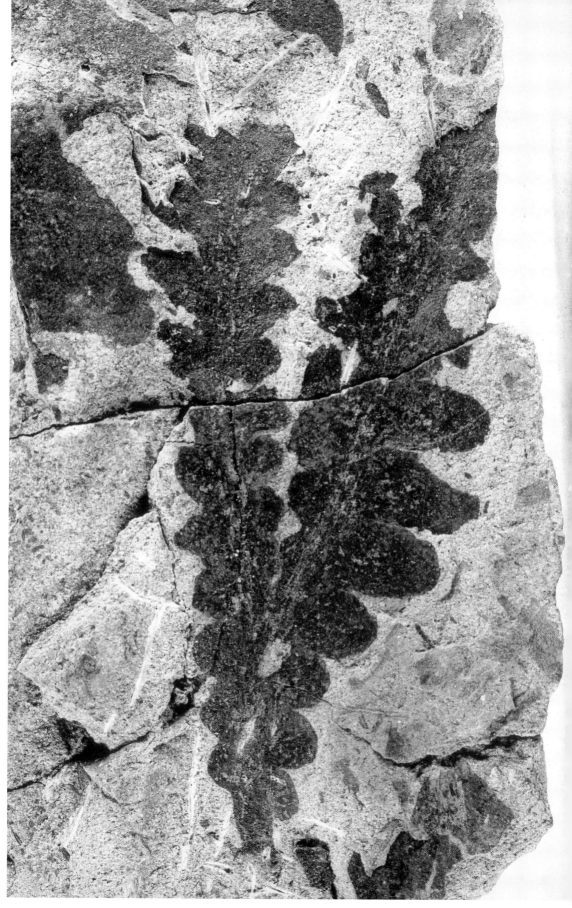

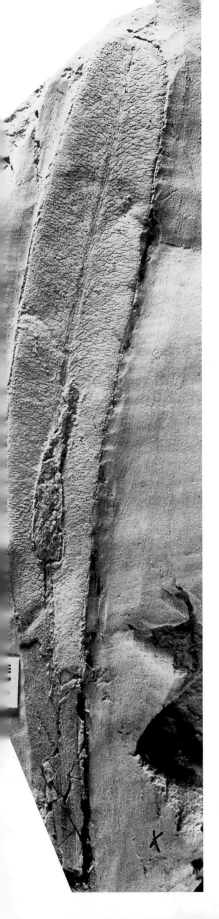

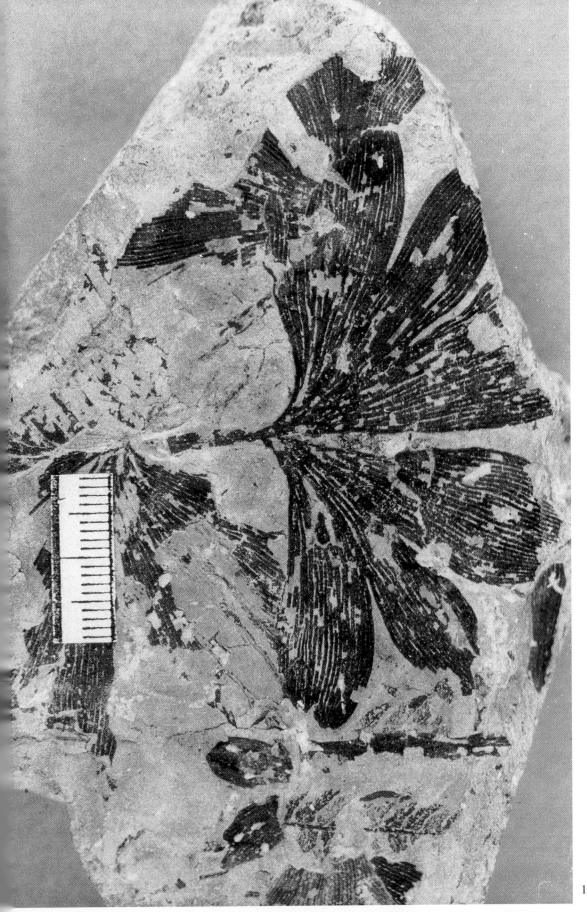

105

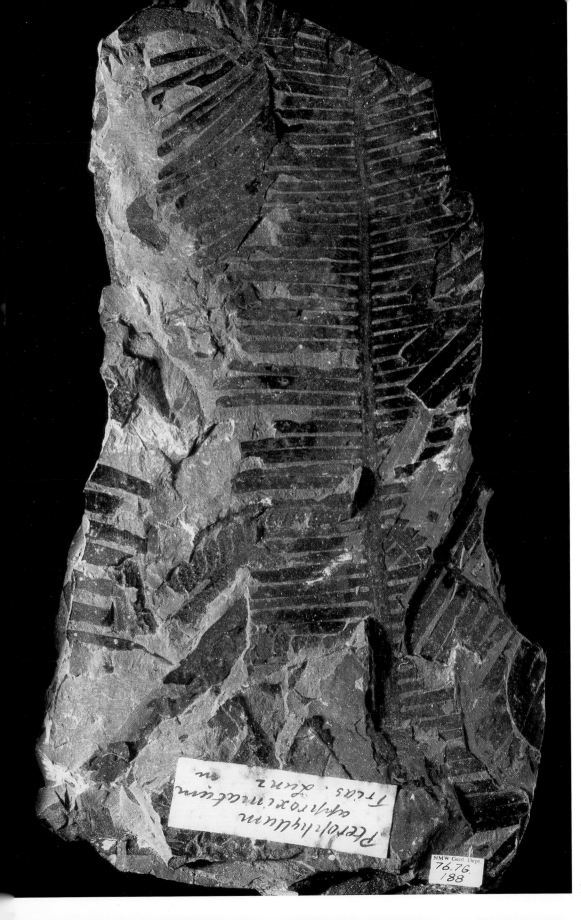

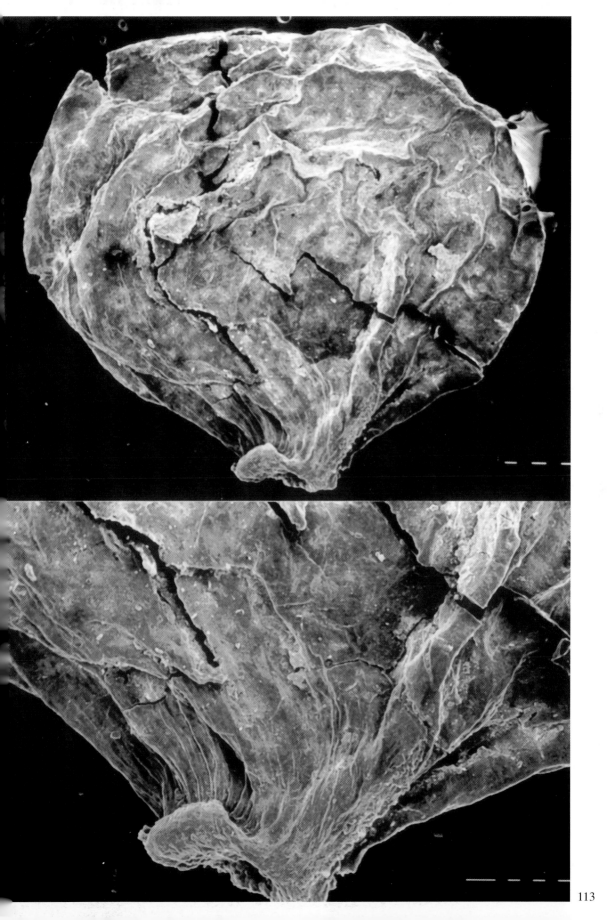

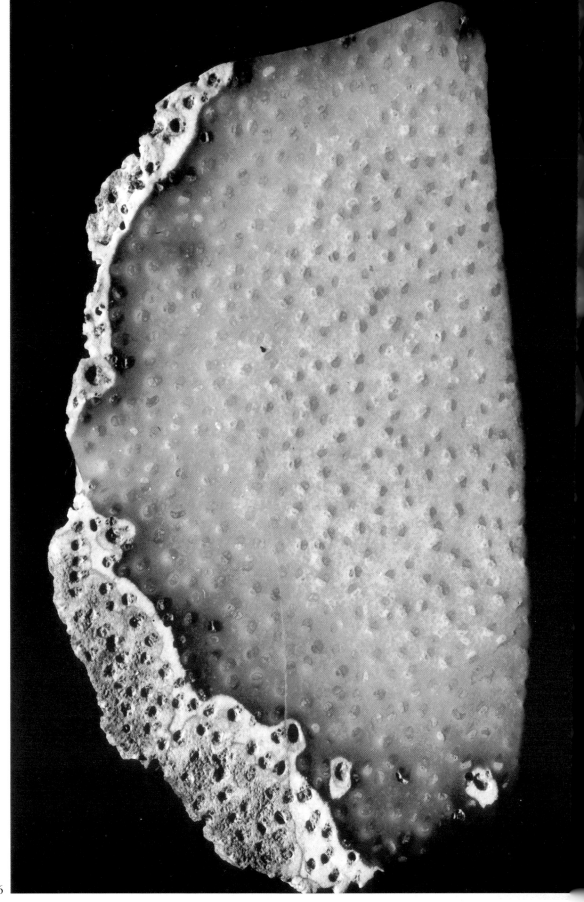

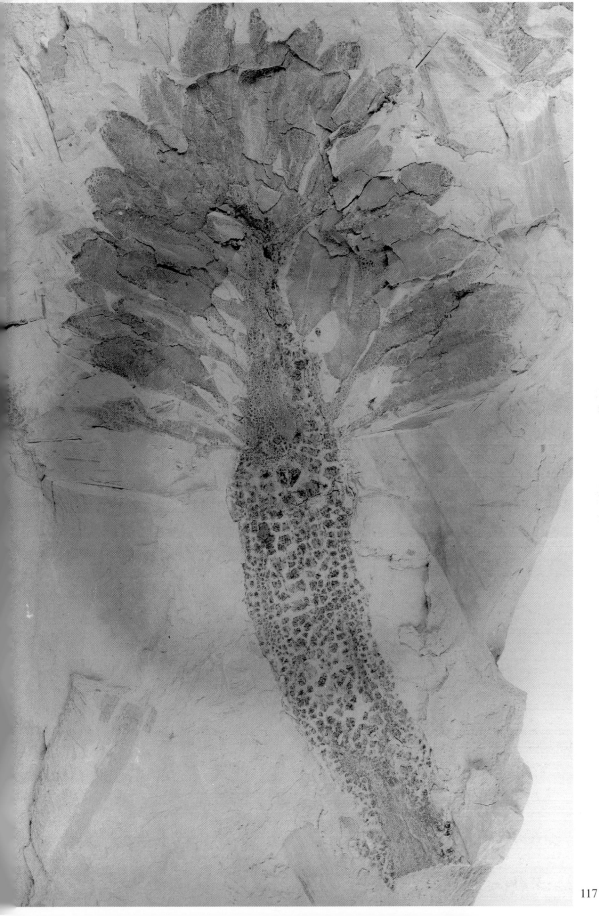

Hazel nuts,
In soil overlying Coed
Madog Quarry, Nantlle.

G. J. Williams' Collection.
27.110 G1337.

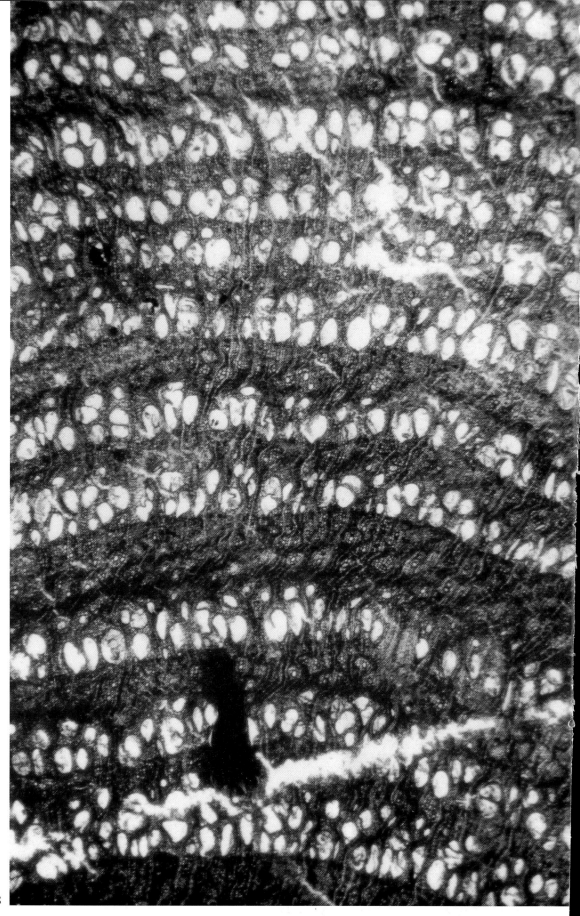